)co.ie

THE COMPLETE
GARDEN
WILDLIFE
BOOK

THE COMPLETE
GARDEN
WILDLIFE
BOOK

Mark Golley

NEW
HOLLAND

For Nadine….
and for Dave, the very dudey cat!

First published in 2006 by New Holland Publishers (UK) Ltd
London • Cape Town • Sydney • Auckland
www.newhollandpublishers.com

Garfield House, 86–88 Edgware Road, London W2 2EA,
United Kingdom

80 McKenzie Street, Cape Town 8001, South Africa

14 Aquatic Drive, Frenchs Forest, NSW 2086, Australia

218 Lake Road, Northcote, Auckland, New Zealand

ISBN 1 84537 210 7

Publishing Manager: Jo Hemmings
Project Editor: Gareth Jones
Editor: Gill Harvey
Designer: Adam Morris
Index: Janet Dudley
Production: Joan Woodroffe

Reproduction by Modern Age Repro Co., Hong Kong
Printed and bound in Malaysia by Times Offset (M) Sdn Bhd

CONTENTS

The Wildlife Trusts

The Wildlife Trusts partnership is the UK's leading voluntary organisation working, since 1912, in all areas of nature conservation. We are fortunate to have the support of more than 343,000 members, including some famous household names.

The Wildlife Trusts protect wildlife for the future by managing in excess of 2,300 nature reserves, ranging from woodlands and peat bogs to heath lands, coastal habitats and wild flower meadows. We campaign tirelessly on behalf of wildlife, even in garden centres. With other leading environmental organisations, we continue the fight to persuade gardeners to boycott peat and limestone pavement – both of which are vital habitats for many threatened species.

We run thousands of events and projects for adults and children across the UK, work to influence industry and government and also advise landowners.

Our Wildlife Gardening initiative encourages people to take action for wildlife in their own back gardens. Readers may have picked up one of our wildlife gardening leaflets or may have seen The Wildlife Trusts' Garden at *BBC Gardeners' World Live*, at the NEC. Visitors to our award-winning show garden included celebrities from the BBC's *Ground Force*, Alan Titchmarsh and Charlie Dimmock.

As traditional wildlife habitats in the countryside come under threat as a result of modern farming techniques, development and water abstraction, gardens are becoming increasingly important. Gardens today are havens for many species of wildlife, providing food, shelter and breeding grounds as well as links to urban parks and other open spaces.

Many of the 46 Wildlife Trusts, which together make up The Wildlife Trusts partnership, employ staff and volunteers to advise people on how best to encourage wildlife to their back yards. London Wildlife Trust even managed to persuade the Prime Minister, Tony Blair, to make space for a child-friendly wildlife pond in the garden at Number 10 Downing Street. It is amazing what a difference a few plants, logs or a pond can make to our wildlife, benefitting species as diverse as the Song Thrush, Painted Lady butterfly, Common Frog or Hedgehog.

The Wildlife Trusts is a registered charity (number 207238). For membership, and other details, please phone The Wildlife Trusts on 0870 0367711, or log on to www.wildlifetrusts.org.

Introduction

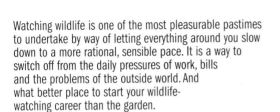

Watching wildlife is one of the most pleasurable pastimes to undertake by way of letting everything around you slow down to a more rational, sensible pace. It is a way to switch off from the daily pressures of work, bills and the problems of the outside world. And what better place to start your wildlife-watching career than the garden.

Wherever you live, there will always be opportunities to enjoy the animals, bugs, birds and insects that pass, often silently, through the garden. Whether your garden is a large, open space in a rural or suburban location or a small 'postage stamp' in the city, there will always be something of interest paying you a visit. Even if you have no garden at all, a window feeder for birds, or a window box to grow plants that attract butterflies, bees and insects, may be all you need to kick start a lifelong passion.

I was very lucky – growing up in a small town in rural west Devon, with a father who had a good basic knowledge of some of the visitors to our very modest garden on the edge of town. At four or five years of age, the birds that came into the garden held my attention for what seemed like hours - engrossed in the colours, the sounds, the different shapes. The sight of a vivid crimson male Bullfinch on the gooseberries has stayed with me to this day. And so do other wildlife encounters from my youth in that small garden - a Hedgehog paying us regular visits one autumn, busily eating the food we'd leave out, summer days spent on hands and knees trying to catch grasshoppers, the first time I saw a Red Admiral gliding past the line full of washing, and the excitement of seeing a huge dragonfly settle on the roses.

Top Right: Male and female Bullfinch *Pyrrhula pyrrhula* (see page 143). **Centre Right:** Red Admiral *Vanessa atalanta* (see page 44). **Bottom Right:** Common Darter *Sympetrum striolatum* (see page 33).

The wildlife that could visit your garden may be so small that you would need a hand-held lens to inspect it properly, or it may be as large as a Red Fox, or even a Roe Deer, when all you need to do is keep as still as possible.

You may want to enhance your enjoyment of watching your garden wildlife by buying a small pair of binoculars. These could prove invaluable when looking at the birds, butterflies or moths that pass by. There are so many different makes and models on the market, offered to you by a myriad of different companies. Seek their advice, and always state the price limit that you have budgeted for! Even some small pairs of binoculars from the leading optical companies can cost several hundred pounds, so make sure you have an idea of just how much you want to pay. There are *always* some good bargains to be had!

This book has aimed to feature as broad a cross-section of visitors as space permits, whether they are mammals, amphibians, moths or birds. Sadly it can only be as comprehensive as the pages allow, and there is a wealth of other species that could pay a visit to your garden.

Top Right: Common Frog *Rana temporaria* (see page 98). **Left:** Great Pond Snail *Lymnaea stagnalis* (see page 17). **Bottom Right:** White-tailed Bumble Bee *Bombus lucorum* (see page 80).

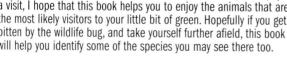

A vagrant moth, butterfly or bird may come into your garden, especially if you are on the coast – rarities can come from almost any point on the compass. Living on the coast of north Norfolk, I'm lucky that my garden bird list now includes Pink-footed Geese and Waxwing from northern Europe, Common Crane and White Stork, off course from southern and eastern Europe and, rarest of all, a Laughing Gull, from North America, which spent much of a summer here in the mid-1990s. To be woken at 5am on a spring morning by the 'sniggering' display call of this charismatic species flying over the house was quite something.

The seasons will have a marked impact on the species that frequent the garden. Many birds head south for the winter, but others come here from northerly climes for the winter. Some species of bird may come to a garden only in the winter as they search for food. Mammals will often make themselves scarce during the winter, as will many of the insect species covered within the book, as well as the reptiles and amphibians. That is not to say that they have left the garden, they may well be nearby, but they are living their lives out of view.

Wherever your garden may be, however many different species pay a visit, I hope that this book helps you to enjoy the animals that are the most likely visitors to your little bit of green. Hopefully if you get bitten by the wildlife bug, and take yourself further afield, this book will help you identify some of the species you may see there too.

Mark Golley
Cley, Norfolk

How to Use This Book

Within this book are a wide variety of species that can be encountered in gardens, of whatever size, across Britain and Ireland. From the most primitive to the most complex, this book aims to give you the key pointers in identifying, and enjoying, the wildlife that may visit your garden.

The book has been arranged with the most primitive invertebrate species coming first, working through to the more complex birds and mammals towards the end. Each group of animals covered in the book has an introductory page, explaining a little about the group that is to follow – general character and behavioural traits, background on the group itself and so on.

As you will read (see pages 12-13), there is an explanation as to how animals are classified. Every species is afforded a scientific name and because the English name may vary in some regions, the scientific name is always stated to avoid too much confusion. Some of the invertebrates do not have common English names, and are, therefore, referred to only by their scientific name.

Each species in the book is illustrated. Alongside the illustration is information on how to identify the animal in question. Also within the text is information on similar species and the habitat, range, behaviour, food and breeding cycle of the species in question. Birds also have notes on songs and calls, important ways to separate some species.

This book aims to enhance your enjoyment of the animals that visit your garden. Identifying the species that visit is tremendous fun and the overall importance of the garden as a wildlife habitat cannot be stressed enough.

The Garden as a Wildlife Habitat

If someone had said, a quarter of a century ago, that the garden would become as vital a wildlife habitat today as the marshes, woodlands, meadows and dune systems that have long been the flagship for wildlife in Britain, they may well have been laughed at. Very few people, even at the turn of the 1980s, would seriously have considered that the garden was any sort of truly viable wildlife habitat, save for a few birds popping in or the odd mouse scurrying around. Thankfully, those far-sighted folk who persevered with their 'garden wildlife' mantra have been proved right, and then some...

Across the globe, habitat is being lost at a frightening daily rate. It is not just the well-publicised 'slash and burn' technique of clearing rainforests. Wetlands are continually being drained. Woodland valleys are lost forever as dam projects take precedence. Heathland and pastures are ploughed up to make way for houses or crops. As the human population increases, business recklessly takes away what is ours. With this in mind, there is perhaps little doubt, now, that the garden is one of the most important wildlife habitats we have.

It is thought that across Britain and Ireland there are well over one million hectares of garden. The sheer mass of habitat, and the variety within, make gardens marvellous refuges for wildlife. But how do you cater for wildlife in a garden? What can you do to make the birds and bees come to you?

A neat, tidy garden where all 'pests' are controlled by pesticides and herbicides and the lawn and flower beds are manicured is likely to be something of a wildlife desert, as is, perhaps a little more surprisingly, a garden that is left to overrun and grow wild. Both types of garden will attract something, but not as much as the garden that is managed specifically for wildlife. All wildlife in a garden needs three essential elements – shelter, food and water. Obviously there are always restrictions given the size of your garden, but even the smallest area can still be managed accordingly.

As shelter, nestboxes serve a dual purpose. In the breeding season it's fingers crossed that a Blue Tit or a House Sparrow will make the most of the sheltered nest site that is on offer. After the breeding season is over, you may be lucky to have Wrens using the box as a wintering site. It is not uncommon to have many Wrens roost together in winter, huddled in a nestbox, providing communal warmth. To ensure that birds nest and 'winter' in any boxes you provide, there must be food, and this, for many species, comes in insect form. These insects also need shelter when they are in go-slow mode. Areas of leaf-litter are ideal spots for insects to hide away, as are compost heaps (which, in turn, provide a warm habitat for hibernating Hedgehogs or torpid Grass Snakes in summer). Another way to provide shelter for insects is by creating a log pile in a warm, sunny spot of the garden. Insects will love this! Usually, as the wood decays and rots down (a process that takes many years), fungi, mosses and lichens may begin to grow, and a whole new mini eco-system will develop.

Birds will come to any garden if there are scraps, breadcrumbs, peanuts or seeds on offer. Bees and butterflies will come if there are shrubs, flowers and bushes. A compost heap will be home to any number of tiny bugs, which creatures like centipedes may feed on. Worms will also love this warm, dank environment and, of course, they in turn become a food source for Robins, Blackbirds, Hedgehogs, Moles, Red Foxes and Badgers.

The last essential in any wildlife garden is water and if you are able to create a pond, so much the better! Even the smallest of ponds is beneficial to the wildlife of the garden. Birds will use the pond to drink and bathe. Mammals, like Hedgehogs, will also use the pond as a mini drinking fountain. If you are lucky, frogs, newts and toads may also decide to use the pond. Larger ponds may also attract dragonflies or damselflies. Try to avoid the goldfish, though – they will eat anything and everything in the pond!

For much more detailed reading on encouraging wildlife into your garden, and managing the habitat accordingly, read John A Burton's excellent *Attracting Wildlife to Your Garden*.

Naming and Classification

The wildlife that visits the garden is a wonderful mass of biology – the complex, the basic, the colourful, the bland. Visitors may be primitive, single-celled organisms (*Protozoa*) through to a developed, intelligent mammal, such as the Red Fox (*Vulpes vulpes*).

This book has specifically aimed itself away from many of the most primitive forms of wildlife that could be found in the garden, simply because many of the *Protozoa* can only be studied properly under a microscope and that is not what this book is about. That being said, it must be remembered that these almost invisible, single-celled organisms play a massive part within the garden environment. This book takes the Silverfish and the humble Common Earthworm as the most primitive species to be found in the garden and works its way through to some of the most highly developed animals in Europe.

Biologists have argued over the centuries about the classification of animals, and many still argue over the minutiae to this day. What is now agreed, and has been in place for many, many years, is the general overview of how animals are now classified. First and foremost is the division into two main groups: animals with backbones and skeletons (the vertebrates) and those without (the invertebrates).

Vertebrates include mammals, birds, amphibians, reptiles and fish (although no fish are included in this book). The invertebrates, such as the Common Earthworm, have no hard parts to their bodies or, like the Common Cockchafer beetle, have a tough outer shell (the exoskeleton) that protects the soft parts of the body.

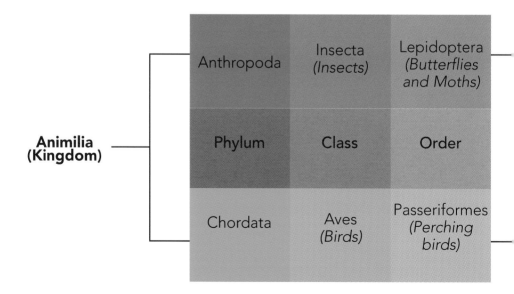

After the first division of animals within the Animal Kingdom (into the vertebrate/invertebrate categories), comes the phylum, which is another very broad, diverse group defined as 'a taxonomic rank, below kingdom, comprising of class or classes'. A class is a specific group of animals, be they insects or birds, as mentioned below. Within each of the classes comes a major, significant division into specific orders: from the class *Aves*, for birds, comes the order *Passeriformes*, meaning perching birds. The next important category is that of family: all the species that share a number of similar characteristics. So, for our example, we know go from the order *Passeriformes* to the family *Turdidae*, which covers all of the world's thrushes. Within each of the family groups comes another division, that of genus, which includes all species that are very similar to each other. Following our example once again, we move from the family group *Turdidae*, to the genus *Turdus*, which includes all so-called true thrushes. Within each genus are the individual species themselves and, within our example, that takes us from the genus *Turdus*, to the species illustrated, the Blackbird (*Turdus merula*). Species are a group that contains all the individuals that have similar characteristics and that can produce fertile young.

Throughout this section, a number of scientific names have been used (as is the case for the rest of the book). These scientific names are based on Latin or Greek and are recognised worldwide. Taking our species example, the Blackbird, the scientific name is *Turdus merula*. The first name refers to the bird's genus, and the second refers to the species itself. Compare the scientific name to the related Song Thrush (*Turdus philomelos*). The close relationship is immediately known with the genus name, and throughout the book, scientific names are used to show where species are related and to aid in their identification.

Family	Genus	Species	
Nymphalidae (Tortoiseshells Fritillaries and Admirals)	Vanessa (Vanessid butterflies)	Vanessa atalanta (Red Admiral)	
Turdidae (Thrushes)	Turdus (True Thrushes)	Turdus merula (Blackbird)	

Chapter One
Invertebrates

The invertebrate group is very varied, with one common characteristic: the absence of a backbone. Several phyla and classes of invertebrates are covered separately in this book: see *Insects*; *Wasps, Bees and Ants*; *Beetles and Spiders*. The most primitive members of the invertebrate group are the single-celled Protozoa; as these creatures can only be seen through a microscope, they have been omitted.

The phylum *Annelida* (of which earthworms are the most familiar members) consists of creatures with segmented bodies, and there are in the region of 7,000 species of worm across the world. The phylum *Mollusca* covers slugs and snails. These creatures are of the class Gastropoda (taken from Ancient Greek, meaning 'stomach foot'). They are soft-bodied and, as the Greek suggests, they move by crawling along on their bellies. The largest phylum within the invertebrate tree is *Arthropoda*. Centipedes and their allies (*Chilopoda*) have anywhere between 20–202 legs, and a tenacious appetite for other invertebrates. Similar in appearance but quite different in behaviour are millipedes (*Diplopoda*): they are vegetarians, feeding below ground on roots and other vegetation. The class *Crustacea* follows. It is perhaps surprising to think of crustaceans in a garden rather than on the seashore, but here they are, present in the guise of the woodlouse. Many *Crustacea* dehydrate in warm, drying air, so hide during the day, only venturing out to feed at night.

Common Earthworm

Lumbricus terrestris Length: 90–300mm; width: up to 5mm

Essential to any successful garden, the Common Earthworm is familiar to all of us. Found across the whole of the country, it likes almost any soil, providing it is not too wet or acid. From this soil comes its nourishment, as it will swallow the earth it lives in and, with an amazing digestive system (for such a 'simple' creature), the Common Earthworm will digest any organic matter its body can find.

Common Earthworms vary hugely in size. Fairly consistent in colour, they are reddish-brown to pink, occasionally with a slight mauve tint to them. The body consists of tiny segments, numbering some 150, with a distinctive brownish 'saddle' along part of its body.

Garden Snail

Helix aspersa Length: 20mm; width: up to 10mm; shell: 25 x 40mm

The Garden Snail is almost certainly the largest snail that will be found in a garden. As well as the environs of a garden, this distinctive snail can be found in woods, parks and waste ground across the country. Its range extends a long way north, meaning that in these cooler climes, the snail must go into a state of near hibernation to avoid the colder conditions.

The Garden Snail has a darkish grey body and its tiny mouth has distinctive white lips. The rounded shell is generally tawny brown with darker flecking, which forms a number of spirals on the shell.

This species feeds on very low vegetation in the garden, often during the hours of darkness. By day, groups of snails communally gather at regular spots, in rather sticky clusters.

White-lipped Snail

Cepaea hortensis Length: up to 15mm; width: up to 10mm; shell: c.14 x 17mm

The White-lipped Snail is a variable, fairly small snail that is found in woods and hedgerows as well as gardens across the whole of Britain and Ireland.

The shell is pale honey brown in colour, with up to five thin dark spirals on the shell. Close up, you can see the feature that gives the snail its name – the white lip at the base of the shell (where it meets the body).

White-lipped Snails feed during both night and day on grass and low vegetation.

Brown-lipped Snail

Cepaea nemoralis Length: up to 15mm; width: up to 10mm; shell: c.18 x 22mm

Brown-lipped Snails are less frequent visitors to gardens than the similar White-lipped Snails, favouring woodland and hedgerows, and their range does not extend as far north.

The Brown-lipped Snail is slightly larger than the White-lipped Snail. The shell colour is extremely variable, ranging from straw to yellow and from brown to pale pinkish. Just as is the case with the White-lipped Snail, the Brown-lipped can show up to five dark brown spiral bands on the shell, but don't be surprised if they show none at all! If the lack of spirals (or otherwise) and the multitude of colours is not confusing enough, the characteristic 'lip' of the shell which gives the snail its name is, indeed, often brown, but it can be extremely pale too. The White-lipped Snail usually has, as the name suggests, a white lip, but this can often be darker, even brown!

This species favours night-time feeding and also feeding after rain, much like other snail species. After a shower, it is not unusual to see several snail species enjoying the damp feeding conditions communally.

Great Pond Snail

Lymnaea stagnalis Length: 30–40mm; width: up to 20mm; shell: 25 x 45mm

For anyone with a pond, the Great Pond Snail is a must to encourage. It favours ponds that are rich in calcium, and can also be found in clean, slow-running rivers, or in canals, across the British Isles.

A very striking creature, the Great Pond Snail has a characteristic shell that is rounded over the body with a distinctive spiral at the rear. The head also has a rather distinctive look, the main antennae being almost ear-like in appearance. The shell colour varies from yellowish brown to dark brown, and always looks very plain.

They are algae feeders and also enjoy chomping away on decaying vegetation. They lay their sausage-shaped eggs on the underside of aquatic plant leaves.

Netted Slug

Deroceras reticulatum Length: 50mm; width: 10mm

The Netted Slug is one of the smaller garden slugs, and is also one of the most widespread and common of all the slugs found across Europe. They are found in fields and pasture as well as gardens.

Netted Slugs are generally pale brown in colour (though they can also be dark grey) with darker flecks across the body. This, combined with rectangular tubercles (the small wart-like parts of the body), give the slug its so-called 'netted' appearance. It also has a rather short-looking rear end.

They eat many different types of plant, but particularly enjoy new seedlings, which, in the eyes of the gardener, undoubtedly make them the enemy! When disturbed, the slug will exude sticky white mucus by way of a defence mechanism.

Great Grey Slug

Limax maximus Length: 200mm; width: up to 20mm

As the scientific name implies, this slug is big! In fact, the Great Grey Slug is the biggest invertebrate that you will encounter in the garden. The species is widespread across Britain and Europe, except in the far north.

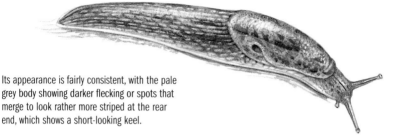

Its appearance is fairly consistent, with the pale grey body showing darker flecking or spots that merge to look rather more striped at the rear end, which shows a short-looking keel.

In gardens they are particularly fond of feeding around compost heaps, where they enjoy rotting plant matter and possibly fungus. They are also found in woodland and along hedgerows.

Like all slugs, Great Grey Slugs are hermaphrodites. When they mate, two individuals slowly climb up a fence, wall or tree trunk before lowering themselves down on a line of sticky mucus. They then go their separate ways before laying their own set of eggs.

Large Black Slug

Arion ater Length: c.150mm; width: up to 15mm

Although another mighty slug, the Large Black Slug is smaller than the huge Great Grey. This species can be seen throughout Europe, and its range extends a long way north, even to colder climes such as Iceland.

The Large Black Slug's name is actually rather misleading, as this particular slug comes in several colour forms other than black – brown, orange or even cream (with a distinctive orange 'fringe' to the body). The back of the body is covered in rather elongated tubercles and there is no keel which gives this slug, in whatever colour form, a distinctive appearance. The further north you go, the darker the slugs tend to be.

The slug is a nocturnal feeder, with a well balanced diet that includes plant material, carrion and dung. As with other slug (and snail) species, it will frequently feed after rain.

18

Garden Slug

Arion hortensis Length: c.40mm; width: 10mm

Compared to other slugs that can be seen in the garden, the Garden Slug is something of a lightweight! The Garden Slug is the smallest family member likely to be encountered in and around the garden (around a centimetre shorter than the largest Netted Slug). Despite its name, it is actually more common on cultivated (agricultural) land, although it will happily munch its way through a lettuce patch or a bed of strawberry plants.

The Garden Slug is generally slate grey to blue-black in colour on the top of the body, with paler flanks and a distinctive orange underside. The colour of its mucus matches the underside, being orange or yellow in tone.

This species of slug is found across Europe, though colder, more northerly climes are not to its liking and it is absent from the northernmost countries of the continent.

Garden Centipede

Scutigerella immaculata Length: c.7mm; width: 5mm

The Garden Centipede is actually not a centipede at all, but a member of another arthropod group, the *Symphyla*. *Symphylans* are closely related to the centipede group, but the common name for *Scutigerella immaculata* has, over the years become, incorrectly, the Garden Centipede. This species is found widely across Britain and Ireland, as well as much of Europe.

The Garden Centipede is small, with around a dozen pairs of rather short legs and a set of long antennae. The favourite habitat in which to try and find this particular invertebrate is around areas of leaf litter and soil, which provide good feeding.

Regarded as a pest in some countries, infestations of this species are common-place; they can decimate areas of newly planted seedlings.

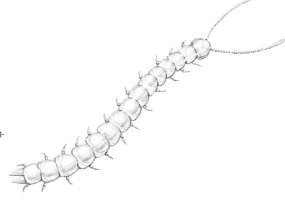

Common Centipede

Lithobius forficatus Length: 20–30 mm; width: c.5mm

The Common Centipede is almost certainly the most frequently encountered centipede in the garden. They are widespread across the whole of Europe and, as well as being a very common garden creature, can be seen in many diverse habitats, from moorland to coastline.

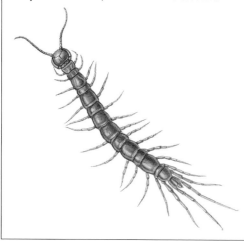

The Common Centipede is twice, or even three times, the length of the Garden Centipede. With an almost ant-like, rounded head, a shiny deep brown body and up to 15 pairs of legs, an adult Common Centipede is a very distinctive sight in the garden. Younger Common Centipedes grow a new set of legs with every 'moult' they undergo, starting with seven pairs of legs when they hatch.

During daylight hours, this species will be found hiding under rocks or logs, but during the hours of darkness, as with other centipede species, they become aggressive and tenacious in their feeding habits, happily taking on slugs, worms, insects and other centipedes as sources of food.

Geophilus carpophagus

Geophilus carpophagus Length: 40mm; width: c.1.5mm

This species of centipede has no common English name, but is a frequent sight in gardens, sheds and cellars, and woodland across Europe.

Rather a long centipede, *Geophilus carpophagus* is long, russet in colour and extremely flexible in the body, having approximately 45-55 pairs of legs. A rather tenacious predator, it is almost always to be found living in soil.

Flat-backed Millipede

Polydesmus angustus Length: 25mm; width: 4mm

A common sight around garden compost heaps, the Flat-backed Millipede is also found in leaf litter, healthy grassland and rich 'loamy' soils across the whole of Europe.

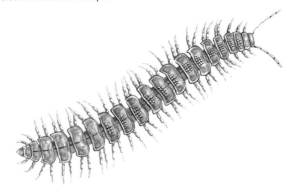

The Flat-backed Millipede is quite centipede-like in appearance, but can be identified as a millipede because it has two pairs of legs per body segment, rather than the single pair per segment that centipedes have. It has up to 37 pairs of legs, and the body segments themselves look rather flattened in appearance. Flat-backed Millipedes are plain mid-brown in colour.

Favourite food items include rotting vegetation, along with roots and occasionally soft fruits such as strawberries.

Pill Millipede

Glomeris marginata Length: 20mm; width: 3mm

The Pill Millipede has a 'hard shell' and some 18 pairs of legs. It is found in a variety of habitats across Europe with gardens being high on the list. They can also be seen in woodland leaf litter, in hedgerows and grassland. Compared to other species of millipede, it is reasonably tolerant of damp conditions.

This species is frequently mistaken for the similar Pill Woodlouse. However, Pill Millipedes are brown, not slate grey, and its plates, particularly the rearmost ones, are of a different shape and structure.

The Pill Millipede eats dead plant material or fresh stems and will, when disturbed, curl up into a tiny ball, just like the Pill Woodlouse.

Common Woodlouse

Oniscus asellus Length: 15mm; width: 5mm

A very familiar sight in the garden, the Common Woodlouse is a distinctive species, familiar to many. They are commonly encountered in rich soils across Europe and almost any logpile left out to rot down over the years will be a home to them. Much the same can be said for compost heaps, which are a haven for the Common Woodlouse too.

The Common Woodlouse is shiny grey in colour with a series of tiny off-white or creamy blotches and edges on the plates that cover the body.

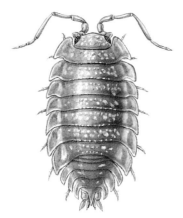

The diet reflects the habitat they surround themselves in, feeding avidly on decaying vegetation and decomposing wood.

Pill Woodlouse

Armadillidium vulgare Length: 20 mm; width: 3–4mm

With a wide distribution throughout Europe, the Pill Woodlouse can be found frequently in gardens, often curled up beneath a wall. It also frequents dry grassland areas (but apparently only those with rich limey soils) and is rather more tolerant of dry conditions than other woodlouse species.

The Pill Woodlouse has a smooth shiny blackish body, which can be tinged with blue, brown or even yellow. Unlike the similar Pill Millipede, the Pill Woodlouse has a more matt look to its dorsal plates, but the main difference is found at the rear end.

On the Pill Woodlouse, the rear is made up of several tiny plates, whereas the Pill Millipede has a single rear-end plate. This is a particularly useful distinction when either species is rolled up into its characteristic ball.

Chapter Two

Insects

The insect class is a huge group of animals totalling one million or more species worldwide, with 100,000 of these being found in Europe. Garden insects come in a vast array of sizes, shapes and colours, but all share certain characteristics.

Adult insects have a body divided into three parts – a head, a thorax and an abdomen. The development, consisting of several stages – from egg to larva to adulthood – is similar, although how and when these stages occur varies within different groups and species to species.

An insect's head has a pair of compound eyes, which are covered in tiny lenses. Some soil-living insects rely on very simplified eyes (ocelli) which detect, rather than see, light. In contrast, dragonflies have many thousands of tiny lenses, enabling them to be active hunters. The head has two antennae that are used for touch and smell, and also contains the mouthparts, which vary according to the creature's feeding methods and just how 'advanced' a species it is.

The thorax is an insect's engine room. It is divided into three segments, with a pair of legs attached to each. The second of an insect's thoracic segments has a pair of wings and if the insect has a second pair of wings, these will be found on the third thoracic segment.

The abdomen contains an insect's sexual organs and is also where digestion and excretion takes place. Females have an ovipositor, which can be very long in some cases.

Silverfish

Lepisma saccharina Length: c.10mm

The Silverfish is a very primitive insect – wingless, with a tapered, scaled, shiny looking body. It has long antennae, and three, even longer, tail-like appendages. Found in houses and garden sheds across Britain and Ireland, Silverfish are often seen scurrying across floors when their cover is disturbed.

The body of the Silverfish is made up of silver-coloured scales and they have very small eyes and simple, biting mouths. The long antennae, tail appendages and legs are all pale in colour.

They are nocturnal feeders, and have a number of choice food items including paper, flour and other starchy items. Apparently, they also enjoy the gum-like glue of modern-day cartons that proliferate on kitchen shelves.

Two-tailed Bristletail

Campodea fragilis Length: 10mm

The Two-tailed Bristletail is found in gardens across the country, with the compost heap a prime location; and although the composition of the heap may be vegetable, Bristletails are actually carnivorous scavengers.

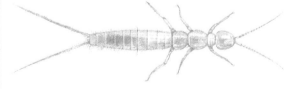

At first glance, the Two-tailed Bristletail is similar to the Silverfish, but a closer look will show a number of differences.

It is around the same length at the Silverfish, but with a slightly different body shape. It is broader with larger scaling, particularly near the head, and the body itself is dark with burnished cream on top of the base colour. The antennae and bristled tail appendages are far longer than those of a Silverfish, and this species has no eyes.

Large Red Damselfly

Pyrrhosoma nymphula Length: 36mm; wingspan: 38–48mm
Flight period: late April to July (occasionally to September in Scotland)

Large Red Damselflies are found across Britain and Ireland, reaching northern isles such as Orkney. They are generally common across their range in Britain, but have declined over recent years in the eastern counties of England, largely due to intensive agricultural practices. The species is common across much of northern Europe, less so in the southern countries of the continent.

The Large Red Damselfly is a neat, mainly red-bodied damselfly and is often one of the first species to be seen in the springtime.

As well as frequenting garden ponds during the breeding season, the Large Red Damselfly can also be seen along the banks of canals, well-vegetated ditches and the acid bogs of heathland. Away from breeding areas, it can be seen along hedgerows and the edge of woodlands, occasionally in large numbers.

Male Large Red Damselflies are pretty unmistakable. The body is almost entirely red, except for black markings towards the rear end (present on the seventh to ninth segments of the body). The head and thorax are black, save for two red stripes on top of the thorax (known as antehumeral stripes). The wings are clear, except for a small dark mark towards the tip of each wing, and the legs are black. When seen from the side, notice the black, yellow and red stripes on the side of the thorax. The female is rather more variable than the male, with differing degrees of black on the abdominal segments.

Azure Damselfly

Coenagrion puella Length: 33mm; wingspan: 36–44mm
Flight period: mid-May to late August

The Azure Damselfly is a widespread species, found right across Britain and Ireland with the exception of northern Scotland and some areas around the Borders and the Wash. Further into continental Europe, the species is seen from southern Scandinavia down to the northernmost areas of Africa.

The Azure Damselfly is marginally larger than the Common Blue Damselfly, which it resembles. The male has a black head and thorax, with blue blobs on the rear of the head, and two thin blue stripes on the top of the thorax. The abdomen is mainly blue with black at the end of each individual segment, except for the wholly blue eighth segment. If you see a specimen close up, look for the distinctive black 'U' shape on the abdominal segment closest to the wing (segment two). From the side, the male Azure Damselfly has narrow blue stripes on the thorax. Young males have lilac replacing the blue, while the female's body appears almost wholly black, with green ends to her segments. There is a blue form of the female, which accounts for around 10% of their number. In both sexes, the legs are black and the wings clear, with a dark mark near the tip of each wing.

Favourite breeding habitats include vegetated garden ponds and ditches, lakesides and slow-moving canals and rivers. Away from their breeding haunts, Azure Damselflies can be seen along hedges with a sunny aspect, field edges and in woodland clearings.

Generally, the Azure Damselfly has a one-year life cycle, but in particularly abundant years or in more northern parts of the range, it can live for two years.

Common Blue Damselfly

Enallagma cyathigerum Length: 32mm; wingspan: 36–42mm
Flight period: mid-May to mid-September

Common Blue Damselflies are found across the whole of Britain and Ireland, with the exception of a small part of south-western Ireland. The species is thought to be the only breeding species of dragonfly on our most northerly island group, the Shetlands, and is probably the most common species of any dragonfly in the country.

As suggested by the name, males are bright blue in colour and are often seen alongside slow-moving rivers, over and around gravel pits, canals and lakes, as well as around garden ponds. Pairs of Common Blue Damselflies are often seen flying in tandem, and the species frequently gathers in loose groups, often numbering many hundreds.

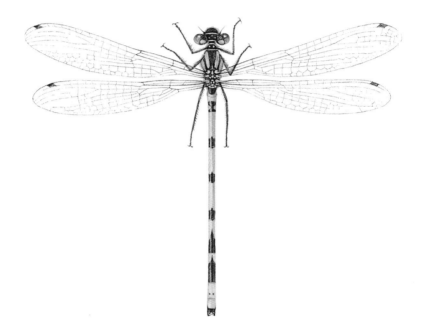

As mentioned opposite, the male Common Blue and Azure Damselflies are rather similar in appearance, the Common Blue Damselfly being slightly smaller in both body length and wingspan. A male Common Blue has a blue and black body, like the Azure, but both its eighth and ninth abdominal segments are completely blue (the Azure has only a wholly blue eighth segment). Also, the black shape on the second segment is 'U'-shaped on the Azure, but a small round blob on the Common Blue. Like the Azure, young males are lilac, but female Common Blues are always blue with black markings – unlike the vast majority of female Azures, which are green and black.

The Common Blue Damselfly is by far the stronger flier of the two species, frequently flying further out over water.

Blue-tailed Damselfly

Ischnura elegans Length: 31mm; wingspan: 30–40mm
Flight period: May to September

Blue-tailed Damselflies are common across Britain and Ireland, although they are absent from some more northerly areas of Scotland. Further afield, the species is common throughout much of central Europe.

The Blue-tailed Damselfly is rather nosy in garden terms, being one of the first species to come and investigate newly dug ponds. It breeds in pools, slow-moving rivers and ponds, and flies low amongst reeds and rushes. It is very variable, with as many as five recognised forms for the female alone.

When seen from above, the male Blue-tail Damselfly has a wholly dark abdomen (black or deep bronze) except for the sky blue eighth segment. The head and thorax are like the abdomen, with small blue markings on both body parts. A side view reveals a blue (or yellow in older males) underside to the segments on the abdomen. Younger males are told by green rather than blue marks on the head and thorax, while newly emerged males (known as tenerals) are buff or lilac toned. Females can appear in many guises – all have black or bronze bodies and all show the blue eighth segment. Some females appear reddish on the thorax; others violet, brown or green; others will actually mimic the male in appearance.

Common Hawker

Aeshna juncea Length: 74mm; wingspan: up to 95mm
Flight period: June to late October

Common Hawkers are indeed common in western and northern areas of Britain and across the whole of Ireland. To the south and east, they become far more localised and are generally found only on heathlands. They are found throughout northern and central Europe.

The Common Hawker is perhaps the largest dragonfly species found in gardens (unless you have a very sizeable pond, which may attract the hugely impressive Emperor Dragonfly, *Anax imperator*). It is a marvellous sight as it flies to and fro with power, speed and grace over still water, whether a pond, a lake or a moorland bog.

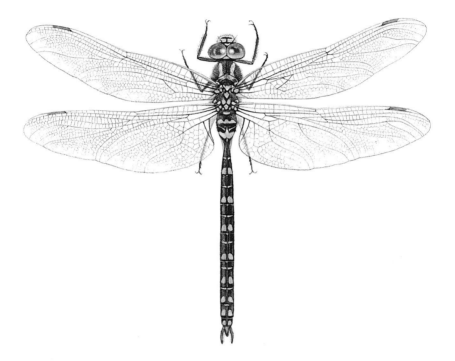

At any distance, a male Common Hawker looks dark, but, close to, the colours are more visible. The abdomen is largely black, with broad blue and narrow yellow marks on each segment. Note too the very tapered appearance at the top of the abdomen, quite unlike any other *Aeshna* dealt with here. The thorax shows narrow yellow antehumeral stripes on black. Female Common Hawkers share the same patterns as the male, but the black of the body is replaced by brown. Both sexes have bold yellow and blue stripes on the side of the thorax.

Male Common Hawkers spend much of their time flying. They seldom rest, and when they do, they settle on branches or low vegetation. A tricky species to sneak up on, the Common Hawker is rather more shy than some other members of the *Aeshna* group - a Migrant Hawker will quite happily come up and buzz you!

Migrant Hawker

Aeshna mixta Length: 63mm; wingspan: up to 87mm
Flight period: July to late October

Formerly just a migrant species from southern Europe, the Migrant Hawker has now established itself as a breeding species in England. It is found in a line running south of the Humber, though it is missing from large parts of Wales. It is also absent from northern England, Scotland and Ireland.

Inquisitive in nature, Migrant Hawkers are happy around slow-running vegetated streams or still ponds. They can also be seen zipping along hedgerows and woodland edges, often in large numbers if conditions are right.

A medium-sized member of the *Aeshna* group (which bears the 'classic' dragonfly shape), a male Migrant Hawker looks predominantly blue and black on the abdomen, while females are browner, sometimes with a dull yellow hue. Immature Migrant Hawkers have grey or very pale pastel blue markings on the black abdomen. A close-up view reveals a tiny yellow triangle at the base of the abdomen (next to the join with the thorax), which differentiates the Migrant Hawker from the Common Hawker. If seen from the side, notice the bold yellow blobs on the thorax.

Southern Hawker

Aeshna cyanea Length: 70mm; wingspan: up to 98mm
Flight period: June to late October

Southern Hawkers are common across lowland areas of England and Wales (with the exception of the very north of Wales) but become more localised further north. They are largely absent from Scotland and almost the whole of Ireland, save for one small area in the south of the country. In Europe, they are common from the Mediterranean to Scandinavia. They breed around garden ponds, lakes, canals and open woodland ponds.

Like all members of the *Aeshna* group, the Southern Hawker is quite large. It is a wonderfully adept flier, colourful, and is undoubtedly the nosiest and bravest of dragonflies in terms of approaching humans. It hovers close to people, generally at waist height, before resuming its patrols along garden edges or woodland glades. Southern Hawkers are also aggressive and frequently attack other dragonfly species.

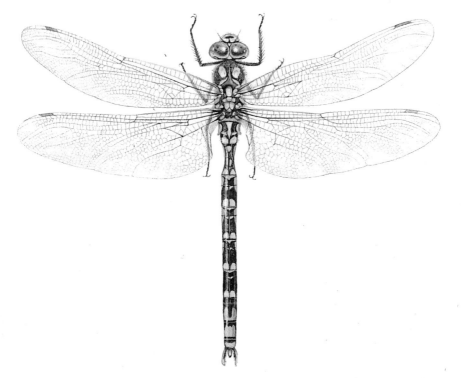

Both sexes resemble the Common and Migrant Hawkers. However, unlike these species, male Southern Hawkers have black abdomens with bold lime green markings, which become bluer towards the lower part of the abdomen (on segments nine and ten, and sometimes on the eighth). Also, the thorax has bold green antehumeral marks and striking greeny-blue slashes of colour on its side. Compared to the male, the female Southern Hawker has a browner abdomen, with paler and greener markings. Both sexes have a broad yellow triangle on the second segment of the abdomen.

Four-spotted Chaser

Libellula quadrimaculata Length: 63mm; wingspan: up to 76mm
Flight period: mid-May to mid-August

The Four-spotted Chaser is a sturdy dragonfly, full of hustle and bustle and rather aggressive in nature, sharing a territorial bent common to many members of the family. Found across almost the whole of Britain and Ireland, its favourite habitat is an open area of calm, still water with plenty of vegetation available to perch on, from which a male will survey his territory – a large garden pond with adjacent reedy vegetation is ideal.

This dragonfly can remain motionless on its favourite perch for minutes at a time before heading out to see off rivals. It is an adept glider, and on summer evenings may gather in small feeding groups, dipping and swooping over water.

Both sexes look similar. The abdomen is broad compared to most dragonflies, and is generally dark brown with small yellow blobs on the sides. Immatures tend to look more orangey-brown than adults. Its Common name indicates the most distinctive feature to look for – the four spots seen singly near the tip of each fore- and hindwing.

Broad-bodied Chaser

Libellula depressa Length: 44mm; wingspan: up to 76mm
Flight period: May to August

The Broad-bodied Chaser is a medium-sized dragonfly and has a chunky, cigar-shaped abdomen. It is less aggressive than the Four-spotted Chaser, favouring shallow ponds or shallow lake margins up to around 1,200m and is found across southern England (into north-western counties) and south and west Wales, but not in Scotland or Ireland.

Broad-bodied Chasers enjoy perching for long periods of time on vegetation next to the water's edge; this is interspersed with spells of fast, direct flight.

The male has a distinctive sky blue abdomen, with small yellow blobs on the side of the body. The thorax and head are dark in colour, and there is a large, dark spot on the base of each wing. The female's abdomen looks grey-brown, while immatures are bright yellow-orange.

Common Darter

Sympetrum striolatum Length: 37mm; wingspan: up to 57mm
Flight period: June to late October (occasionally into November and, rarely, December)

The Common Darter is a familiar member of the dragonfly family, found throughout Britain and Ireland apart from in some areas of Scotland. Favoured habitats are ponds, lakes and brackish lagoons up to around 1,800m. Common Darters can often be seen some way from water, resting on twigs and branches.

Unlike most species of dragonfly, the Common Darter frequently lands on the ground, particularly in cool conditions, and returns constantly to the same spot. Common Darters are also seen in quite large numbers as the year progresses, flying around busily in the late autumnal sun.

The male Common Darter has a deep orange-red abdomen, a double patch of yellow on the thorax and clear wings (except for a buff patch at the base of his four wings). Females and immatures are yellowish-green to light brown in colour. Compare the shape of the body with the very similar Ruddy Darter (see below).

Ruddy Darter

Sympetrum sanguineum Length: 34mm; wingspan: up to 55mm
Flight period: June to October

The Ruddy Darter is very similar to the Common Darter in size, shape and colouration. As with the Common Darter, during its flight period, it can be seen around vegetated ponds, lakes and canals, and frequently around the edge of woodland. It also enjoys perching on overhanging twigs. The range of the Ruddy Darter is, however, quite different. Whereas the Common Darter is found across most of the country, the Ruddy Darter is missing from the whole of northern England and Scotland, much of Wales and the north and west of Ireland.

To the very practised eye, it is possible to distinguish between the two dragonflies in flight. Ruddy Darters have a rather more 'bouncy' flight than the Common Darter, and seem to hover a little longer.

The male is bright crimson red in colour with dark (rather than yellow) sides to the thorax. The females and immatures are rich ochre in tone, though the immatures are slightly more golden. With a really good view, you can see that the legs are black, as opposed to the Common Darter's greyish blue. Perhaps the most striking feature of the Ruddy Darter is the distinctive shape of its abdomen, which is tapered and narrows along the body like an old-fashioned beer pump handle.

Speckled Bush-cricket

Leptophyes punctatissima Length: male 10–15mm, female 20–25mm

The Speckled Bush-cricket is a small but distinctive insect, one of 10 members of the bush-cricket family found in Britain. It favours garden undergrowth, parks and woodlands across Britain to southern Scandinavia and the Mediterranean. This species and the others within the *Tettigoniidae* family were formerly known as long-horned grasshoppers because of their amazingly long, fine antennae. Many grasshopper experts are now happier with the name Bush-cricket as these insects are more closely related to true crickets than they are to grasshoppers.

Aside from its incredible antennae, the Speckled Bush-cricket's most obvious feature is its very long hind legs. The sexes are similar in appearance – 'new leaf' green all over, with fine red speckles. The male has a narrow brown line running from the tip of the head to the end of the abdomen, while the long ovipositor of the female should be visible too.

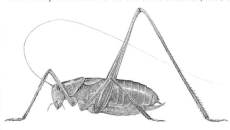

Unusually for Bush-crickets, both sexes sing, the male's song being stronger and more resonant than the female's. It is a series of 'zzt' sounds, repeated at short intervals. Once attracted to a male and his song, the female will reply to alert him to her presence. Adults are seen from July to October, feeding on leaves of garden plants such as rose bushes or raspberries.

House Cricket

Acheta domesticus Length: up to 20mm

The House Cricket is a native of Africa and the Middle East, but it has now become established across the whole of Europe, including Britain. It favours the warmer parts of a house or garden, particularly the kitchen, but is more common around commercial premises such as bakeries. Another prime location is the rubbish dump – the warmth of decomposing refuse is right up the House Cricket's street!

House Crickets are true crickets (defined by having the forewing laid horizontally over the body, rather than vertically as in Bush-crickets), and one of just three species seen in Britain. The two native species, the Field Cricket and the Wood Cricket, are both rare and confined to the southernmost counties of England. It is omnivorous, but its food of choice is decaying vegetable matter.

The House Cricket's body is generally brown or sometimes yellow, with black on the head and wings. The wings extend beyond the tip of the abdomen. Females have an ovipositor that is almost as long as the insect itself. It has long, rather hairy back legs and very long antennae.

The song of the this species is a soft, gentle warble, heard at dusk or at night.

Common Field Grasshopper

Chorthippus brunneus Length: male c.15mm, female nearly 20mm

The Common Field Grasshopper is a relatively small grasshopper found widely in dry, grassy areas across much of Britain, as well as northern and central Europe. In this country, it is particularly widespread in the southern counties of England.

Adults are seen during the later months of summer, from July to October. They vary widely in colour from olive green, grey-green, black to even dull purple, though generally it is the dark olive green form that is most commonly encountered. Both sexes have narrow wings, which always extend beyond the tip of the abdomen.

The tip of the male's abdomen is reddish, as is that of some females too. The song, a hard sounding 'zzzt', is repeated every two seconds or so.

Common Earwig

Forficula auricularia Body length: c.10mm; pincer length: male 10mm, female c.5mm

The Common Earwig is very common across Britain, Ireland and the rest of Europe. It is found in many varied locations, and frequently in gardens (even very small ones), where they may be seen around old logs or heaps of garden waste. They are nocturnal and look for dark crevices during the daytime (gaps in a wall are ideal).

The Common Earwig is one of those unfortunate, misunderstood insects that sends shudders down the spines of many people. Folklore has decreed that they will seek out an ear and then burrow inside to break through the eardrum. This is, of course, nonsense, though anyone who has enjoyed a camping holiday may well have encountered one in the middle of the night! The male Common Earwig has rather fearsome-looking pincers and can give you a sharp nip.

Both sexes are similar in colour, the pale brown head and thorax contrasting with a mahogany abdomen, while the legs and antennae are straw coloured.

Interestingly, the female Common Earwig shows a great deal of maternal care for her young. After laying between 20–50 eggs in soil, she looks after them during the winter, then feeds and cares for the young until they are fully developed and can leave their nest.

35

Green Shield Bug

Palomena prasina Length: 10–15mm

Seen from spring to autumn, the Green Shield Bug is fairly common across Britain and Europe. It is found mainly along woodland edges and clearings, but it also favours gardens with a variety of shrubs and herbaceous borders. These garden plants act as the prime food source for the bug.

The Green Shield Bug is a small but instantly recognisable insect due to its colour and shape. It is bright green all over except for its darker brown wingtips.

As the seasons progress, the Green Shield Bug becomes rather more bronzed, resembling the Hawthorn Shield Bug (*Acanthosoma haemorrhoidale*), but note the difference in size and shape. It hibernates in leaf litter over the winter.

The Hawthorn Shield Bug is a little narrower and longer in the body than its green cousin, and distinctly bronze-toned in colour throughout the year. As the name suggests, the Hawthorn is a great place to see the species, along with gardens and hedgerows.

Common Pond Skater

Gerris lacustris Length: c.10mm

Being small and wholly blackish-brown, the Common Pond Skater appears rather primitive, but it is one of the most cleverly developed insects covered in this book. Common Pond Skaters are found across Britain and Ireland, as well as Europe. This species is one of nine types of Pond Skater found in Britain.

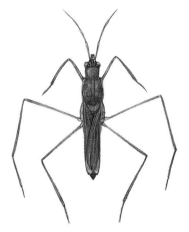

The Common Pond Skater is likely to be seen on the surface of almost any area of calm water, from a garden pond to a slow-running river. It has a broad body and a short head, complete with large, rounded eyes (quite unlike the Water Measurer - see opposite). It is often fully winged, and flies away from water prior to hibernation.

On the water, Common Pond Skaters are in their element. Each set of legs is used to maximum effect - the hindlegs are used as a pair of rudders as the insect powers across the water surface using its middle set of legs. The front legs are then used to catch tiny insects on the water surface.

Water Measurer

Hydrometra stagnorum Length: c.10mm

The Water Measurer is common and widespread in this country. Found on the water surface of garden ponds, slow-moving rivers, canals and lakes, the Water Measurer is easily told from the Common Pond Skater by its thin body and amazingly long head.

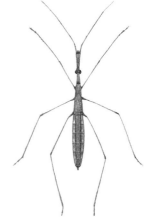

The Water Measurer is a very thin surface-dwelling insect, with a narrow body, long, thin legs and a massively elongated head. It is dark grey-brown all over, with paler panels on the side of the abdomen.

The Water Measurer feeds on the water surface but, courtesy of the elongated head, it will actually 'spear' its prey, be it a Water Flea or a larvae, through the surface, holding on to it with its snout (correctly known as the rostrum). This method of feeding suggests that the elongated head is instrumental in the process.

Water Scorpion

Nepa cinerea Length: c.20mm; 'tail' length: 10mm

The Water Scorpion is a tenacious and fearsome predator, found in garden ponds with shallow edges or in the margins of lakes and quiet rivers. Water Scorpions are found in suitable habitats across Britain and Ireland, and throughout Europe.

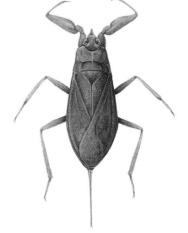

The Water Scorpion is rather flat in appearance, and the insect is entirely olive green to pale green brown in colour. The 'tail' at the rear of the abdomen is actually a siphon, used as an intricate form of breathing apparatus when the Water Scorpion is below water.

Although fully winged, it rarely flies, and it is not a great swimmer. It feeds on tadpoles, small insects, even tiny fish – its nasty looking front legs are excellent tools to catch prey with. The insect can be seen throughout the year.

Green Lacewing

Chrysopa pallens Length: c.20mm; wingspan: 30–40mm

The Green Lacewing is a familiar sight on warm summer and early autumn nights, frequently finding its way into houses via open windows. On the wing between May and late August, the Green Lacewing is found in gardens, woods, hedges and any well-vegetated area across much of Britain and Ireland, though it is absent from Scotland.

With its sizeable length, long antennae and a wingspan that is almost double the body length, the Green Lacewing is quite unmistakable. The body is bright lime green, with golden compound eyes. The wings are transparent, but covered in thin green veins. At rest they are held 'roof-like' over the soft body.

A friend of the gardener, it preys on many different aphids, with Greenfly very high on its list of favourite foods.

Flea

Cat Flea – *Ctenocephalides felis* Length: up to 2mm
Human Flea – *Pulex irritans* Length: up to 3–4mm

The Cat Flea (**below left**) has a scientific name that is almost longer than the creature itself. This rather ugly, brown or blackish insect is commonly found in houses where cats and dogs are kept, although they will visit a larger number of hosts than you might imagine.

Along with the 60 or so species of flea found in Britain and Ireland, the Cat Flea is a minute, wingless insect with a curious 'flattened on either side' look. All flea species are ectoparasites with mouthparts specifically developed for sucking blood.

Fleas are rapid breeders in warmer weather, and infestations can be a real problem for the animal lover, as well as the pets themselves. Cat Fleas and Dog Fleas often swap around, visiting alternate canine or feline hosts, and they can cause irritation to humans.

Just like the Cat Flea, the Human Flea (**above right**) is rather flat on the sides of the body. It is also an ectoparasite. As well as living and feeding on humans, the Human Flea is found on animals such as the Badger, the Red Fox and other hole-living mammals. The ascent to humans is thought to have occurred in the age of cave people. Interestingly, the Human Flea is now far less common than it was a century ago.

Thunder-fly

Aeolothrips intermedius Length: up to 2mm

The Thunder-fly (colloquially known as 'thrips', from its scientific name) is a common sight (and feeling!) across Britain and Ireland during the summer months, especially for those living in areas of dense vegetation, or adjacent to arable farmland where crops such as wheat are grown.

This insect is minute and black, with two pairs of narrow wings. Along with the 150 or so species of 'thrips' that also occur across Britain and Ireland, the Thunder-fly can be seen feeding on the underside of numerous flower species, and it particularly enjoys the sap of plants such as Dandelions and Yellow Crucifer.

In good years, these creatures may drive you mad by them crawling over you! They don't bite, but their incessant movement around your head and body can become infuriating. They can also be a menace inside the house, creeping behind picture frames and appearing as tiny black marks under the glass.

Large White

Pieris brassicae Forewing: 25–35mm Flight period: March to late October

Common across Britain, Ireland and Europe, the Large White butterfly is a menace as far as the gardener is concerned. All the hard work put into the vegetable patch can be seriously undone if a few Large White caterpillars start to munch your brassicas.

Mostly creamy white above, both male and females have bold black tips to the forewings, with small blackish-grey spots on both the fore- and hindwings. The spots on the female are rather bolder than on the male; the spots on the male's forewing can be rather diffuse. The female Large White butterfly has buttery yellow hindwings. Both sexes show buttery yellow tips to the underside of the forewing and wholly butter yellow undersides to the hind-wing. The sexes share the two black spots visible on the underside of the forewing.

Large White caterpillars are green, with clear black spots and yellow stripes. The number of broods each year depends on weather conditions. They are found in many different habitats, but gardens with lots of plant life will always be visited by the Large White.

Small White

Artogeia rapae Forewing: 15–30mm Flight period: early March to November

The Small White is a butterfly equally adept at causing vegetable patch havoc as its larger relative, the Large White. With two to four broods a season, their abundance in the gardens of Britain and Ireland, as well as the rest of Europe, makes them a virtual pest species.

The Small White is almost a miniature mirror image of the Large White. Both sexes are creamy white on the upperwings, with black suffused tips to the forewing and spotting on both the fore- and hindwing. A male generally has just one spot on the forewing, the female two; overall, the female's upperside looks dingy compared to the male's. Both the male and female Small White have butter yellow tips to the underside of the forewing, and wholly yellow undersides to the hindwing. The underside of the forewing on both sexes has small black spots.

The caterpillars of the Small White butterfly are green bodied with a bold yellow stripe along the flank.

Orange Tip

Anthocharis cardamines Forewing: 20–25mm Flight period: April to late June

The Orange Tip is one of the first butterfly sentinels of spring and is on the wing from the first warm days of April. Relatively common across Britain and Ireland, but absent from the northern parts of Scotland, they can be seen along woodland glades, hedgerows and meadows as well as gardens, and overwinter in pupa form.

Around the same size as the Small White, the male and female Orange Tips are readily distinguishable. The male sports bold, broad orange patches on its forewings, and a good view shows neat black wingtips lying over the orange. The inner part of the forewing is cream with a single black spot. The female lacks the orange of the male; black tips and a single black spot contrast with the white remainder of the forewing. Both sexes are mottled grey-green on the hindwing. The sexes look similar on the underside - the hindwings are marked with moss green marbling, while the underside of the forewing is black and white on the female; orange, black and white on the male.

Favourite food plants include Honesty and Sweet Rocket in the garden, and wild plants such as Lady's Smock and Garlic Mustard.

Brimstone

Gonepteryx rhamni Forewing: 25–30mm Flight period: early March to May
(for hibernated individuals); otherwise June to July

The Brimstone butterfly is common across the whole of Britain and Ireland, with the exception of the north of Scotland, and along with the Orange Tip, this species is a herald of spring.

The colour of the male makes this rather large butterfly unmistakable. It is bright sulphur yellow on the whole of the upperside, save for single red spots on each wing. The underside is pale leaf green in tone, with a hint of yellow on the forewing. A female Brimstone is pale cream with a hint of lime, and lacks any of the black marks seen on the similarly sized Large White. Like the male, she has individual red spots on each wing (seen on both the upper- and underwing). The Brimstone's shape is also distinctive – a very rounded fore- and hindwing with markedly tapered, almost pointed, wingtips.

The larvae are green, with strong white stripes along the flanks, and feed on plants such as buckthorn. Hibernating adults take to holly or ivy for the winter.

Small Copper

Lycaena phlaes Forewing: 10–17mm Flight period: May to early October

The Small Copper is indeed a small but very bright and distinctive butterfly. Although the species is found across Britain, Ireland and Europe, it is not the commonest visitor to the garden, though on a nice warm sunny day you may be lucky.

Male and female Small Coppers are similar in appearance. The forewing is almost wholly bright copper in colour with a chocolate brown border. Within the copper are several bold black spots. The hindwing is largely brown with a copper-coloured line of markings near the tip, the copper encased in black. From below, the forewing retains the copper colour but not the intensity of tone, while the brown wingtips are paler too. The black spots remain prominent. The underside of the hindwing is pale brown, with black spotting and a copper-orange border on the rear of the wing.

Small Coppers may produce two to three broods a year; the later broods can be smaller and more intense in colour. The small green caterpillars feed on Common and Sheep's Sorrel, and dock leaves.

Common Blue

Polyommatus icarus Forewing: 14–18mm Flight period: May to October

The Common Blue is the scarcer of the two Blue butterflies that are likely to venture to the garden. A species that enjoys well-flowered grasslands, roadside verges, dunes and even waste ground, the Common Blue will however pop into the right garden at the right time.

The male Common Blue is, at first glance, similar in appearance to the Holly Blue. This species tends to be a little larger (though not always) and the male is a deeper shade of violet blue than the male Holly Blue. With practice, the difference is easily detected. Unlike the Holly Blue, a male Common will show no black at the tip of the forewing. From below, the male Common Blue will have grey-brown fore- and hindwings, with small rows of black spots (circled in white) and, towards the end of the hindwing only, some orange edged spots. Female Common Blues are actually brown, and therefore very different from the male. The whole of the upperside is a deep woody brown, with orange dots near the edge of the forewing, and a similar set of dots, tipped with black and white near the edge of the hindwing. The underwings of the female are similar to the male, but with more extensive orange markings near the edge of both fore- and hindwings.

Across some parts of Britain, and throughout Europe, there can be much variation within Common Blue populations. Some females are deep metallic blue in colour, while others are dark brown, with very broad orange bands on the wings. Some males can mirror the pattern of the female on the underside, while others can be very deep shades of violet.

The Common Blue will have two or three broods during the year, and some of these can be rather dark as the year progresses. Favourite food plants are mainly vetches and other legumes. The caterpillars are small and green, and they overwinter in caterpillar form.

Holly Blue

Celastrina argiolus Forewing: 12–18mm Flight period: early April to June;
July to early September

The Holly Blue is another butterfly that takes to the wing in early spring. The species is common in England and Wales, but absent from large swathes of northern England and the whole of Scotland, and is only sporadically found in Ireland.

The entire upperside of the male is a beautiful bright violet blue, with a narrow black and white chequered border to the fore- and hindwings. The underside of the male is powder blue, with a small amount of delicate black spotting on both sets of wings. Again, the fore- and hindwings have a thin chequered border. The female is less intense in colour, and has an obvious black border to the tips of the forewing and the inner part of the hindwing. This pattern is mirrored on the hindwings, where the overall colour is powder blue, like the male.

The caterpillar is small, green and almost slug-like in appearance. It feeds on a variety of plants, and by adulthood is seen on holly and other fruiting trees. Adults are also known to drink sap, honeydew and the juices from decaying carrion. Holly Blues overwinter in pupa form.

Peacock

Inachis io Forewing: 20–25mm Flight period: March to May (hibernated individuals);
otherwise June to September

The Peacock is one of the most instantly recognisable species of butterfly that grace the garden. Common throughout the whole of Britain and Ireland, and across much of Europe, the Peacock is often seen within the warm confines of the house during the winter. These are the individuals seen, post-hibernation, in March and April.

Peacocks are sizeable butterflies, with broad wings that appear a little ragged at first glance, but are actually very neatly shaped and contoured. The sexes are nearly identical in appearance. The 'base' colour of the upperside of the butterfly is maroon, with thin black edges to the wing and a beautiful 'eyed' look on each individual wing – violet, yellow, white, black and maroon suffusing to resemble the 'eye' of a male Peacock feather. From below, the butterfly may look drab and dark, but closer inspection will reveal wonderfully intricate patterns on the blackish grey-brown underwings.

The caterpillar is chunky, rather bristly, and feeds vigorously on nettles. During the summer, this is one of the species that benefits from the garden buddleia bush.

43

Red Admiral

Vanessa atalanta Forewing: 30mm Flight period: March to April (hibernated individuals); otherwise June to October

Along with the Peacock butterfly, the gorgeous Red Admiral is one of the most easily identified and most instantly recognisable species seen in Britain's gardens. Found widely across the whole of Britain, Ireland and Europe in almost any area of flowering open space, its numbers vary from year to year in this country, depending on the numbers of Red Admirals that move northwards from southern Europe.

A sizeable species, males and females look virtually identical. The forewings show broad black tips with white spots, contrasting with a broad pillar-box red band across the middle of the wing. The rest of the forewing is deep brown in colour, reminiscent of crushed velvet. The hindwing is predominately the same velvety brown, except for broad red tips dotted with black. On the underwing, the forewing is incredibly complex in its markings. The most prominent feature is the red band, edged with black. White blobs are also obvious near the tip, and thin vibrant blue lines sit just above the red. The hindwing is a wonderful, vaguely psychedelic, fusion of brown, grey, black and even blue.

Red Admiral caterpillars are very dark and bristly with a yellow stripe along the flank. There are two broods during the course of the year. As with the Peacock, some Red Admirals hibernate inside houses during the winter, emerging on warm March and April days.

Painted Lady

Vanessa cardui Forewing: 30mm Flight period: May to October

The Painted Lady is an attractive migrant butterfly species that is found across much of Britain during the late spring, summer and autumn, but is relatively scarce in Scotland and Ireland. In some years, literally hundreds of thousands of Painted Ladies have been seen flying into the country along the coastlines of eastern and southern England.

This species is slightly larger than the Small Tortoiseshell butterfly, but similar in shape. The upperside markings are delicately intricate – the fore-wing has bold and extensive black tips, with small white blobs. The rest of the forewing and the whole of the hindwing is warm orange-brown, browner near the body, with neat black patches. The rear of the hindwing shows a neat row of small black dots, circled by orange. The underside of the wing is a mass of colour and patterning. The forewing shows a swathe of orange, with black and white running through and around it. The tips of the wing are pale grey-brown, with small white dots. The underside of the hindwing is a mass of variegated pale browns, cream and white, with a row of four blue and black 'eyes' near the tip of the wing.

Painted Lady butterflies can be seen in any flower-rich area, be it a garden, roadside verge, woodland glade or hedgerow. They are very fond of buddleias and use thistles and nettles as host plants for their eggs (they have two broods during a summer here). The caterpillars are black, with neat tufts of bristles and a distinctive yellow and red stripe along the flanks.

Small Tortoiseshell

Aglais urticae Forewing: 25mm Flight period: March to April (hibernated individuals); otherwise May to October

The Small Tortoiseshell is the butterfly that you are most likely to encounter during the winter months, close to a window in the house, fluttering at the sunshine outside. The species is widespread and very common across Britain and Ireland.

On average, the Small Tortoiseshell is slightly smaller than a Painted Lady. The sexes are almost identical in appearance. The upperside is largely bright orange-brown, with black and yellow patches along the leading edge of the forewing and a brown area across the inner half of the hindwing. Both the fore- and hindwings show a neat row of black-rimmed blue spots close to the trailing edges. From below, the forewing is a pale, dull yellow, reminiscent of very old newspaper. The tip of the forewing is mottled grey-brown and a black patch is clear along the front of the wing. The hindwing looks largely velvety black in colour, with grey and buff fine lines extending to the row of blue spots near the tip.

This species, like several others, will festoon a buddleia bush during the summer. The Small Tortoiseshell can be seen almost anywhere with an abundance of flowers, from the garden to roadsides to woodlands and waste ground. The black, bristly caterpillars are often seen feeding on nettles.

Comma

Polygonum c-album Forewing: 23mm Flight period: March to April (hibernated individuals); otherwise late May to September

At the turn of the 20th century, the Comma butterfly declined rapidly across much of its range in Britain, and this slump lasted for nearly three decades. Thankfully, from the end of the 1930s, numbers began to rise, and now this vibrant, beautiful butterfly is again common across England and Wales (but absent, as it always was, from Scotland and Ireland).

The Comma's most obvious feature is the amazing, ragged look to its wings. The forewing is deep orange, with black spots and blotches and a dark brown edge. The hindwing is browner along the inner edge, with the rest being orange and black, much like the forewing. The underside is largely dark blackish-brown, with a paler area near the tip of the forewing. Look for the little white comma-like mark on the underside of the hindwing, which gives the butterfly its name. Both sexes are almost identical in size and markings.

Commas are seen in gardens (they are fond of buddleias), flowering hedgerows and particularly at the edge of woodlands. Their caterpillars look a little like bird droppings - black, with a white tip at the end. These caterpillars feed on nettles, elms and, in south-eastern counties, hops. There are two broods during the year, the second brood always being darker in colour than the first. As the autumn fades, adults look for a suitable wintering site amongst the remaining leaves.

Meadow Brown

Maniola jurinta Forewing: 20–26mm Flight period: May to September

Meadow Brown butterflies are common across Britain and Ireland, favouring areas of open grassland and woodland edges. There is some regional variation in appearance (some darker, some more orange).

Whilst not the most glamorous of butterflies, on closer inspection, the Meadow Brown is not quite as dowdy as it first appears. The sexes are slightly different, mainly in size – the female is always larger than the male. Both have largely pale brown fore- and hind-wings when seen from above. The inner areas of the forewing are rather more orange in tone than the rest of the butterfly and, note too, the double black spot with white 'highlight' on the forewing. On the underside, the orange on the forewing is very prominent, contrasting with the narrow brown edge. The black spot is clearly seen. The hindwing appears largely pale grey-brown, with a cream bar running from top to bottom on the outer third of the wing.

The species winters in larval form. The caterpillars are green, with a white band across the flanks, and feed on tall grasses.

Gatekeeper

Pyronia tithonus Forewing: 17–25mm Flight period: July to September

The Gatekeeper butterfly is common across its rather limited range within Britain and Ireland. Found widely across most of England and Wales, it is absent from more northern counties of England and Scotland and is present only in a narrow band across the southern tip of Ireland.

Superficially resembling the larger Meadow Brown, a good view will show a number of differences. The forewing is brighter orange, with a more obvious brown border. Like the Meadow Brown, the Gatekeeper has a bold double black spot on the forewing, but in this species they are both dotted with white. The hindwing tends to be a brighter orange and brown than that of the Meadow Brown, and shows a single small black spot. The underwing mirrors that of the Meadow Brown, but note the double black spots on the forewing and more variegated grey-brown patterning on the hindwing.

Gatekeepers are found along almost any hedgerow and in many gardens. They are also frequently seen in damp woodlands. The caterpillar is usually green, but can be brown, and feeds largely in bramble patches and other dense areas of vegetation or shrubbery.

Speckled Wood

Pararge aegeria Forewing: 19–22mm Flight period: late March to early June; late June to October

One of the earliest butterflies to take to the wing in spring, the Speckled Wood is instantly recognisable. It has become increasingly common, but has a patchy distribution across Britain and is largely absent from some eastern and northern counties in England and some parts of Scotland. It is found across the whole of Ireland.

There is little difference between the sexes, although females are generally a little darker than males. The upperside of a Speckled Wood is entirely dark brown; cream spots pepper the top of the forewing, and form a neat row on the edge of the hindwing. The underside has a variegated cream and black look to the forewing, with a single black and white spot near the tip. The hindwing is less well marked, but shows a row of tiny black dots near the tip of the wing.

The Speckled Wood is common along woodland rides and clearings, but is also frequently seen in gardens, especially those adjacent to wooded areas. The green caterpillar feeds on grasses, and the species overwinters in both larval and pupal forms.

Wall Brown

Lasiommata megera Forewing: 20–26mm Flight period: April to October

The Wall Brown butterfly is common across its range in this country and, as the name may suggest, is often seen basking on walls. In Britain, it is found throughout much of England and Wales, but is absent from more northern counties of England, the whole of Scotland and Ireland. Male Wall Browns are notoriously furtive and are very easily disturbed from footpaths, rocks and walls.

A medium sized butterfly, the upperparts are intricately marked. The general colouration is a bright, rather deep orange, criss-crossed with neat brown patterning. The forewing shows a single 'eye' (black with white centre), while the hindwing shows a row of up to four 'eyed' dots. The underside of the forewing is similar to the upperside, with brown markings scoring the orange base colour. The hindwing is variegated grey and black, with a row of dark 'eyes' near the wing tip. Female Wall Browns tend to be slightly larger than males.

Wall Brown butterflies favour areas of rough pasture and adjacent gardens. Their green caterpillars feed in grass, and the species has two or three broods during the year. The species overwinters in larval form.

Common Clothes Moth

Tineola biselliela Forewing: 4–6mm Flight period: year round

The Common Clothes Moth may only be small, but it is known as the most destructive of all the Clothes Moth species. Almost always seen indoors rather than out, the species is very common, and is found throughout Britain and Ireland.

The Common Clothes Moth is small and generally pale all over. The forewings are a delicate very pale yellow colour, contrasting with creamy or sometimes silvery hindwings. The antennae are relatively long for a moth of this size, while the body is very pale cream.

An adult Common Clothes Moth actually flies very rarely, much preferring a hurried crawl when disturbed.

The white-bodied, brown-headed larvae will eat animal fibres, and are known to make tiny shelters from them too.

Green Oak Roller

Tortrix viridana Forewing: 10mm Flight period: May to August

The wonderfully named Green Oak Roller moth (also known as the Green Oak Tortrix) is common across the country, and found in parks and gardens as well as woodland. Whilst it is perhaps not a species that you will recognise immediately, there is every chance that you have encountered it when walking through deciduous woodland and dodged the green larvae hanging from a tree by a fine thread. These green, centimetre-long caterpillars are the larvae of the Green Oak Roller.

The adult, when emerged, has lovely fresh emerald green forewings, which contrast with cream hindwings. The body shows a pale lime cast to it and the antennae are rather small and fine.

After emerging, the lifespan of these night-fliers is painfully short, probably no more than a week. The larvae feed on buds and the unfurling leaves of oak trees.

Gold Fringe

Hypsopygia costalis Forewing: 8mm Flight period: July to October

The Gold Fringe moth is a rather small, delicate-looking moth, found only across southern England.

Both the fore- and hindwings show a similar pattern – the base colour of both wings is milk chocolate brown with a gold fringe, as the name suggests. The forewing has two bold golden spots on its leading edge, while the hindwing has two narrow golden lines running down across the wing from the golden spots. The body is brown and the antennae shortish. When seen from below, the hindwings are purple tinged, but retain the gold fringe.

Gold Fringe moths are found in hedgerows near pasture, some gardens and grassy areas, and fly at night. The larvae are white with a brown head; they feed on dead grass and even roof thatch.

White Plume

Pterophorus pentadactyla Forewing: 12–15mm Flight period: May to August

The White Plume is a rather amazing looking moth. Despite its small size, its ghostly white colour and amazing feathery 'split' wings make it unmistakable. Given its fondness for bindweed, the White Plume moth is relatively common in gardens, but can also be seen on waste ground and along hedgerows across Britain and Ireland.

The whole moth is white in colour, including the body and the antennae. The forewing is split into what can only be described as feathery sections, while the two hindwings are split into three more. Note how long the legs are compared to other moths of a similar size.

White Plume larvae are bright green in colour, complete with tiny tufts of silvery hair. They feed on bindweeds, and will spend the winter as a small caterpillar.

51

Small Magpie

Eurrhypara hortulata Forewing: 15mm Flight period: June to August

Small Magpie moths are intricately and beautifully marked moths that can be found widely across Britain and Ireland, in gardens with nettle patches, hedges, woods or waste ground.

The base colour of the Small Magpie's wings is silvery-grey, with broad grey and black borders to both sets of wings and neat, intricate markings in marbled black elsewhere. The black and grey markings on the wings are in marked contrast to the yellow body, dotted black.

The caterpillar is green and feeds almost entirely on stinging nettles or related plants. They will overwinter in a cocoon spun among plant debris.

Leopard Moth

Zeuzera pyrina Forewing: 20–35mm Flight period: June to August

A wonderful looking moth, the Leopard Moth is a resident species found in most southern counties of England and the east of Wales. It is absent from northern Britain and Ireland, where it has been recorded just once.

Leopard Moths have creamy white wings with fine black spots (heavier and denser on the forewing), a furry thorax with six bold, black spots, and a series of greyish rings around the broad, cigar-like abdomen. The male is notably smaller than the female, which can measure up to 35mm across. Note the feathered, or comb-like, silver-grey antennae.

The Leopard Moth caterpillars are creamy with black spotting and a dark head. They feed on a number of different trees, such as Willow, Cherry, Hawthorn and Beech, by tunnelling into the tree itself, where they pupate.

Leopard Moths are found in gardens, parks, woods and orchards. They are readily attracted to light at night, and can sometimes be seen resting on tree trunks during the day.

Garden Carpet

Xanthorhoe fluctuata Forewing: 14mm Flight period: April to October

The Garden Carpet moth is variable in appearance, ranging from silvery-grey to almost black. The species is commonly found in gardens with vegetable patches, on allotments or rough ground, as well as in woodland and open coastal areas. They may be seen resting on walls and fences during daylight hours.

The forewings are silvery-grey, with variable amounts of darker grey markings along the leading edge of the wing and towards the tip. There will always be a small, dark triangular patch on the wings where they meet the thorax. The underwings tend to show less in the way of darker marks, generally being wholly pale, with fine grey lines across the wing. The dark thorax, usually grey, will always contrast with the white (black-dotted) abdomen.

Garden Carpet caterpillars are also variable, ranging from grey-green to dark brown. They strike a familiar 'looping' pose and look like tiny twigs. They will feed on brassicas, nasturtiums and wild plants such as Garlic Mustard and Hairy Bittercress. They may produce up to three broods in a year.

Small Emerald

Hemistola chrysoprasaria Forewing: 18mm Flight period: May to August

The Small Emerald is a distinctive moth with a beautiful, subtle green colouration and rather rounded forewings. The species is not the commonest of garden moths, with a rather patchy distribution across the country. It is found most frequently in southern England and Wales, with some reaching the north-west, whilst it is largely absent from Scotland and Ireland.

Small Emeralds are pale emerald green all over, except for two white lines running across the forewing and one across the hindwing. The body is also pale emerald green, except for a brownish tip to the abdomen.

The caterpillars are pale green too, and feed on cultivated Clematis and Traveller's-joy.

Small Emeralds are found largely on downland, woodland fringes and areas of scrub as well as in some gardens.

Lackey Moth

Malacosoma neustria Forewing: 13–20mm Flight period: June to August

The Lackey Moth is a common species, found across much of England as far north as Yorkshire, southern and western counties of Ireland and lowland and coastal areas of Wales. It is far scarcer in Scotland. It favours sunny areas in gardens, open areas of woodland, scrub and hedges.

Lackey Moths range from reddish-brown to yellowy-brown. The rounded wing shape is quite distinctive; the forewing is largely plain except for a broad dark bar across the middle, and is bordered by paler lines. Note the brown and white chequered pattern on the trailing edge of the forewing. The hindwings are similarly patterned (without the checks), and are generally paler in colour. The body is tawny-brown and furry in appearance and the antennae slightly feathered.

Lackey Moth caterpillars are quite long, with hairy tufts along the length of the body. The body itself is blue-grey with multi-coloured stripes (white, black, orange and red) along the flanks. They live in communal cocoons and feed on Hawthorn, Blackthorn and Sallow. The species has one brood in a year and winters in egg form.

Peppered Moth

Biston betularia Forewing: 20–30mm Flight period: May to August

The Peppered Moth is another beautiful moth, with incredibly intricate markings, particularly on the forewing. Found across much of Britain, in woodlands, hedgerows, scrub, parkland and gardens, the Peppered Moth has managed to adapt its colour to its surroundings in some areas of the country.

The white form of the Peppered Moth is the most common. The base colour of both fore- and hindwings is white, with the forewing intricately patterned with black, especially on the front edge. The hindwing is also intricately marked with black, but shows more white, especially near the body. The antennae are feathered and whitish, while the furry body shows grey-black markings on both the thorax and abdomen.

A sooty-black form also exists. This was particularly common in more polluted areas, particularly in the 1980s. In London, for instance, the sooty form accounted for between 60–80% of the population (*Waring et al.*), and in some areas the percentage was even higher, relating directly to atmospheric pollution. There is an intermediate form, too, which is common in less polluted areas.

The caterpillar is another 'looping' caterpillar. Green or brown in colour, they can reach up to 60mm in length. Peppered Moths feed on a wide variety of trees and shrubs, including Sallow, Raspberry, Hawthorn and Sweet Chestnut.

Winter Moth

Operophtera brumata Forewing: 15mm Flight period: October to January

On winter nights, a car journey is barely complete without a pale moth being caught, momentarily, in the headlights. This, and any that may appear around the window of your house during the winter months, will almost certainly be a Winter Moth. They are very common and are found right across the country, including the northern isles of Orkney and Shetland.

The forewing is generally grey-brown with darker patterning reminiscent of the age rings on a tree trunk. The most noticeable mark is a dark central band. The hindwing is largely plain, pale cream to very pale brown in colour.

The eggs are laid in trees, and the green, looping caterpillars emerge in spring. In summer, they fall to the ground where they pupate prior to winter emergence. The caterpillars feed on deciduous trees and are a pest species to fruit farmers.

The flight period of the Winter Moth coincides with that of the November Moth and, in more northern localities, the scarcer Northern Winter Moth (a rather silky, paler-looking moth). The flying moths are all males; females are flightless.

Magpie Moth

Abraxas grossulariata Forewing: 20mm Flight period: June to August

The Magpie Moth is one of several black and white moths of a similar shape, but its relatively large size makes it unmistakable. Whilst variable in patterning, the species' colours are generally consistent. Widespread across England, Wales and Ireland, with a more sporadic distribution across Scotland, Magpie Moths are seen in gardens, allotments and hedgerows, but are commonest in northern moorland areas.

Magpie Moths have black and white spotted forewings; either colour can dominate, from almost all black to almost all white. Generally, however, spotting is the norm, as is the yellow on the centre and base of the forewing. The hindwing is usually much whiter than the forewing, with just a few black spots present, particularly near the tip of the wing. The body is deep yellow, with a black mark on the thorax and small spots on the abdomen.

The caterpillar is up to 30mm long, pale green with dark spots and has a rust-coloured line on the sides. Chosen food plants include Blackthorn, currant bushes (Red and Black), Privet, Gooseberry and Hazel.

August Thorn

Ennomos quercinaria Forewing: 17mm Flight period: August to September

August Thorn moths are widespread but localised across England and Wales (they are commonest in western counties), extending north to Yorkshire and Cumbria. They are largely absent from Scotland, but are again fairly common across Ireland. August Thorns are found in woods, parks, some downland, and gardens too.

The August Thorn is a very neat moth, with a distinctive tapered look to both fore- and hindwings. The jagged appearance of its wings when closed is rather reminiscent of the Comma butterfly.

The colour of the species varies from orangey-yellow (usually the male) to pale straw yellow (the female). The forewing has two narrow brown lines running across the wing, and sometimes a brown border too. The hindwing is largely plain, except for a varying degree of dark speckling. The antennae are buff to golden, feathered in appearance, and the body is furry.

The larva is grey-brown and another 'looping' caterpillar. With tiny nodules on the body, it is twig-like. It feeds on Oak and Elm trees.

Hummingbird Hawkmoth

Macroglossum stellatarum Forewing: 25mm Flight period: April to December

The Hummingbird Hawkmoth is a day-flying, migrant species to Britain and Ireland, seen in varying numbers each year – in some years, thousands may arrive, while, in others, very few may appear. The species gets its name from its hummingbird-like feeding behaviour – it visits assorted tubular flowers in search of nectar, which it takes in through its long proboscis.

This is a relatively small Hawkmoth with buff-brown forewings that have two narrow black lines running across them. The hindwing and the underside of the wing, which are both obvious in flight, are deep orange. The head and thorax are grey and furry, while the abdomen is grey with black and white checks on the flanks.

Hummingbird Hawkmoths feed on many plants, but favour Viper's Bugloss, Buddleia, Red Valerian and various kind of Clematis. The species tends to appear here during late summer and early autumn; spring records are likely to relate to hibernating individuals, though it is suspected to be a breeding species in the far southwest of England. The caterpillars are green with a band of yellow, white and green flank stripes. They feed on Lady's Bedstraw and Hedge Bedstraw.

Given its feeding preferences, the Hummingbird Hawkmoth can be seen almost anywhere across the country, in gardens, woodlands and along the coast.

Eyed Hawkmoth

Smerinthus ocellatus Forewing: up to 45mm Flight period: May to July

This large and very beautiful hawkmoth can be seen in gardens and places where willow predominates, such as scrubland, river edges and woodland. Eyed Hawkmoths are relatively common and are found across England as far north as Cumbria, Wales and Ireland, but are absent from Scotland. Its combination of size, wing shape and colouration make it one of the most distinctive moth (and indeed insect) species covered in this book.

The large forewings of the Eyed Hawkmoth are wavy along the trailing edge, with an intricate, variegated pattern of buffs, greys and creams. The pattern rarely differs, but the colours can range from pinkish-brown to dark chocolate or even sooty black. The underwing is pinkish near the base of the body, fading to grey towards the tip. The most startling feature is the large blue 'eye' on each wing, encircled with black, designed to deter passing birds who fancy a tasty meal. Eyed Hawkmoths will, when disturbed, gently wiggle to and fro, exposing the eyes on the wing.

Eyed Hawkmoth caterpillars are just as distinctive as their parents. Large and bright green in colour, they show seven yellow diagonal stripes on the side of the body and a blue-green 'horn' at the rear end. They feed on apple trees and willows.

Poplar Hawkmoth

Laothoe populi Forewing: 40mm Flight period: May to September

The Poplar Hawkmoth is another large and unmistakable species. It is perhaps most striking at rest, when the wavy edges of both fore- and hindwings are obvious and the hindwings can be seen to project out in front of the forewing. The forewings themselves are arced and narrow, leaving a gap between the wing and body.

The patterns on both sets of wings are very consistent. The forewing has a broad dark bar across the middle, on which there is a small silver comma mark. There is often a dark area along the trailing edge of the forewing as well. The paler patches (usually grey) are variegated in appearance. The hindwing is less patterned, generally grey, with a bold cinnamon orange patch at the base of the wing. The colouration is variable, from grey to pinky-brown.

Up to 60mm in length, the caterpillars are green with seven diagonal yellow stripes (like the Eyed Hawkmoth) but with a yellow rear end 'horn'. They feed in poplars and sallows.

The Poplar Hawkmoth is the most common and widespread Hawkmoth in Britain and Ireland, even reaching the Inner Hebrides. It is seen in gardens, woods, heathland or parks, or any area where the caterpillar food plants grow close by.

Privet Hawkmoth

Sphinx ligustri Forewing: 55mm Flight period: June to July

This is a *monster* of a moth! Some individuals can be nearly 60mm across when in flight, and even at rest, they look pretty big. Its size, combined with the patterning on the abdomen and the wings, make this moth totally unmistakable. Privet Hawkmoths are quite common in England, south of a line from the Severn to the Wash; further north, records are less frequent, and the species is absent from much of Scotland.

The long, broad forewings are greyish brown to buff in colour, with thin black 'veins' running across them. A thin white and black line is also apparent near the tip. The hindwings are usually pinkish to honey in colour, with three black bands running across them from leading edge to trailing edge, and a pale pink base. Perhaps the most striking feature of all, however, is the boldly striped abdomen, shocking pink contrasting with black, although the intensity of colour can vary. The thorax is blackish-grey and furry.

Adults feed on nectar, enjoying honeysuckles in particular. The caterpillars are green with seven purple and white stripes along the side of the body. Favourite food plants include Wild and Garden Privet, Guelder Rose and Lilac. The species winters in pupal form underground, sometimes to a depth of a foot or more. Newly emerged Privet Hawkmoths can often be seen resting on fence posts or tree trunks.

Yellow-tail

Eupractis similis Forewing: 20mm Flight period: June to August

Yellow-tail moths are common across much of England and Wales, but largely absent from Scotland. They are widespread in Ireland, though not overly common. They can be seen in gardens, hedgerows, woodland and areas of scrub.

A rather unassuming moth, the Yellow-tail actually hides a rather unpleasant secret. It is not the moth itself that provides the irritating surprise but the caterpillar: its hairs can cause a nasty rash if they come into contact with skin. Up to 40mm in length, the Yellow-tail caterpillar is black with red stripes and white spots, and feeds on fruit trees, Hawthorn and Blackthorn.

The moth itself is wholly white, the body included, with the exception of the abdomen's yellow rear and small dark grey marks along the trailing edge of the forewing (on males only). The body is very hairy and the antennae are feathered on the male, but not on the female.

Buff-tip

Phalera bucephala Forewing: 30mm Flight period: May to August

A truly exquisite and unmistakable moth, the Buff-tip is, at rest, one of the most distinctive moths to visit the garden. Commonly found across England and Wales, they are less frequent and more localised in Scotland. They are widespread in Ireland.

The Buff-tip's forewing is a delicate silver-grey crossed with narrow black and cream lines. At the tip is a bold buff-yellow to orangey mark which, when the moth is at rest, bears a remarkable resemblance to a broken twig. The hindwing is entirely cream. The thorax is greyish with a buff mark, contrasting with a grey-brown abdomen.

The caterpillar is up to 40mm in length, blackish, with thin white lines along the length of and circling the body. It is covered in numerous tufts of hair. It feeds on deciduous trees, including various oaks, elms, Beech and Hornbeam. The caterpillar pupates and winters in pupal form below ground.

Buff-tips can be seen in gardens, woodlands, parks and areas of scrub.

Common Footman

Eilema lurideola Forewing: 15mm Flight period: June to August

A rather small but characteristic moth, the Common Footman is found across England and Wales, but becomes much less frequent the further north you go. The species is widespread in Ireland.

The forewing is wholly lead-grey in colour, except for a narrow (but obvious) yellow border to the leading edge and tip of the wing – particularly noticeable when the moth is resting. The hindwing is pale lemon yellow, with a very narrow dark trailing edge. The body is grey, with a small yellow tip to the abdomen.

Common Footman caterpillars are grey and hairy, with black lines along their backs and red lines on their flanks. They feed on Hawthorn leaves, Bramble, Traveller's-joy and tree lichens. The species can be seen in many lowland areas, from urban gardens, farmland and wetlands to coastal sites.

Buff Ermine

Spilosoma lutea Forewing: 15–20mm Flight period: May to August

The Buff Ermine is a medium-sized, attractive moth that is found widely across England, Wales and Ireland, but is less frequent in Scotland.

The forewings are butter yellow, sometimes paler to almost cream in colour, with dark smears near the body and the trailing and leading edge of the wing. The dark smears and lines vary between individuals. The hindwing is entirely butter yellow, with perhaps a single grey mark near the leading edge of the wing. The thorax is greyer than the abdomen, which shows some dark flecks on the side.

Buff Ermine caterpillars are pale grey and covered in a mass of long brown haired tufts. The caterpillars favour heathers and other wild or garden herbaceous plants, such as nettles and Honeysuckle.

White Ermine

Spilosoma lubricipeda Forewing: 15–20 mm Flight period: May to August

The White Ermine moth is similar in shape and size to its family relative, the Buff Ermine, but is more widespread and can be seen across Britain and Ireland.

The forewing of the White Ermine is white with varying amounts of black spotting, which is most prominent near the tip of the wing. The hindwing is also white with black spotting, but with generally far fewer spots. The thorax is white and furry, while the abdomen is yellow with black spots on the sides and centre.

In more northern parts of Britain, and also in Ireland, the White Ermine can look rather more creamy or yellowish. In Scotland it may even look brown.

The caterpillars are dark brown and, like the Buff Ermine, very hairy. They feed on docks, nettles and other herbaceous plants in gardens, grasslands, woodlands, moorland and coastal sites.

Garden Tiger

Arctica caja Forewing: 25–35mm Flight period: June to August

The Garden Tiger moth tends to favour open locations such as meadows, scrub or open woodlands, but is a frequent visitor to gardens. Though found across much of Britain and Ireland, sadly, it has seen a steady decrease in numbers across much of its range, probably because of increasingly mild mid-winters followed by harsh cold snaps later on, which kill the overwintering caterpillars.

With its wings closed, the Garden Tiger is utterly unmistakable. The forewings close over the body and the chocolate brown and white spots, dots and blotches form beautiful patterns. When the moth opens its wings and reveals the hindwings, the contrast is amazing – bold, rich orange with big black blobs. The body is thick, with a brown thorax and orange abdomen striped with black, while the feathered antennae are whitish.

The caterpillar is almost as distinctive as the adult – it is a long, very dark and incredibly hairy beast, known as a 'woolly bear'. The larvae often spend time basking in warm sun and they feed on many different herbaceous plants, including nettles, docks, Hound's-tongue and many garden species.

Setaceous Hebrew Character

Xestia c-nigrum Forewing: 20mm Flight period: May to October

The Setaceous Hebrew Character moth is found widely across Britain and Ireland, favouring cultivated areas (including gardens) as well as lowland woods, meadows and marshes. Its remarkable name is perhaps its most striking feature.

The name is derived from a black, vaguely saddle-shaped mark on the central portion of the forewing, which is said to resemble a letter of the Hebrew alphabet. This mark contrasts strongly with the pale buff patch on the centre of the forewing, which extends to the leading edge. The remainder of the wing is highly patterned in grey, black and buff. The underwing is rather plain compared to the forewing – wholly pale cream, with a whiter trailing edge. The thorax and abdomen are furry and brownish in colour.

The species always has two broods a year, and overwinters as either a caterpillar or a pupa. The caterpillar is green, but as it gets older it becomes grey toned. It feeds on willowherbs, nettles and burdocks.

Green Arches

Anaplectoides prasina Forewing: 20mm Flight period: mid-June to mid-July

The Green Arches is a rather beautiful and unmistakable moth. Found commonly throughout Britain and Ireland (except for the northern isles), the Green Arches is seen primarily in broad-leaved woodland, but also in gardens.

A newly emerged Green Arches is a wonderful, intricately patterned moth, with several shades of green on the forewing. Dark greens, lime greens and olives merge together in patterns with dark black edges. The hindwing is generally dark olive, palest at the base, darkest towards the tip, but encased by a narrow, pale border. The thorax and abdomen are both furry and olive in tone, the thorax being paler than the abdomen.

The caterpillar is brown with darker smudges and feeds on Bramble, Honeysuckle, docks and Primrose. For this reason, the moth frequents broad-leaved woodland and gardens where these plants grow.

Green Arches overwinter in larva form from August to May, hiding in leaf-litter by day and feeding at night. They pupate below ground.

Large Yellow Underwing

Noctua pronuba Forewing: 25mm Flight period: June to October

It is not unknown for people running large moth traps, particularly at coastal sites in late summer, to catch literally thousands of this species in a single night. They can be seen in almost any habitat and are found widely across Britain and Ireland, with a population of both resident and migrant moths.

Large Yellow Underwings have long but narrow forewings. They vary in colour from bright buff-brown to dark chocolate brown. The forewings of the male are variably coloured and marked but there is always a small black dot near the wingtip along the leading edge. Females tend to be less varied in colour and are less marbled in appearance. The hindwing on both sexes is always a deep, bright yellow with a strong, broad black bar near the trailing edge of the wing. The yellow colouration is thought to act as a deterrent to birds.

The caterpillars are green, with a dark head and two rows of black spots along the back of their body. They can measure up to 50mm in length. The caterpillars feed on a wide variety of herbaceous plants including docks, Marigolds, Foxgloves and brassicas.

Burnished Brass

Diachrisia chrysitis Forewing: 20mm Flight period: May to October

The Burnished Brass moth is found widely across much of mainland Britain and Ireland. The species favours gardens, areas of rough grassland, woodlands, hedgerows and marshes.

The species takes its name from the bold metallic markings on or near the tip and at the base of the forewing, which can vary from brassy gold to brassy green in colour. Between the metallic markings is a broad brownish patch running across the wing. The broad, rather hooked forewings are held above the body when the moth is at rest. The hindwing is basically pale, with darker (usually brown) areas across much of the wing that get darker towards the tip. The body is rather furry, with a distinctive tuft on the ginger-toned thorax. The abdomen is grey-brown.

Burnished Brass caterpillars are blue-green to green in colour, with a white line along the flanks and thin white stripes around the body. They feed on nettles, Wild Marjoram and many other herbaceous plants. The adults have a fondness for feeding at dusk on Red Valerian, Honeysuckle and other plants rich in sweet nectar.

Silver Y

Autographa gamma Forewing: 20mm Flight period: May to September
(but can be seen year round)

The Silver Y is a very common and widespread moth, which is an annual migrant species to Britain and Ireland. Arriving from May onwards, the numbers each year vary widely, but in warm summers the numbers of these day-flying moths can be huge. Silver Y moths that arrive here in early part of the summer will also breed here.

The forewing is variable in colour, ranging from very dark grey-black to brownish. The wings are patterned and variegated, and a small silver 'Y' is always visible on the middle of the forewing. The hindwing is usually brownish in tone, palest (almost grey) at the base, darkest towards the trailing edge of the wing. The body is furry and grey.

Silver Y caterpillars are green, up to 20mm or more in length and feed on a wide variety of low-growing wild or cultivated plants, from bedstraws and nettles to cabbages and beans.

This species flies during the day, and can often be seen feeding alongside butterflies on buddleia bushes, often in some numbers. In fact it can be seen almost anywhere as it searches for nectar– any sweet-scented plants in a garden, particularly near the coast, will be sure of a visit.

63

Angle Shades

Phlogophora meticulosa Forewing: 25mm Flight period: May to October and at other times of the year too

The Angle Shades is a medium-sized moth with a unique shape when at rest. The species is both resident and migratory, with a widespread distribution across much of Britain and Ireland. It is a frequent visitor to gardens.

At rest, the first thing you will notice about an Angle Shades is the ragged, cut-out look to the rear of the forewing, and the way in which it holds it wings in a kind of crease. The forewings are generally olive green with pinkish to pinkish-brown inverted 'V' marks near the leading edge. The pattern is very consistent, but the colours can vary. The hindwing is generally pale buff in colour, darker towards the trailing edge, with a dark border.

The big green caterpillar feeds on a selection of wild and garden cultivated plants, including Red Valerian, nettles, dock and Bramble.

Angle Shades can often be seen resting on walls, fences or on garden plants during daylight hours. They can be seen almost year round, but are most common in late summer when continental immigrants swell resident populations.

Red Underwing

Catocala nupta Forewing: 30–35mm Flight period: August to September

The Red Underwing is a very large moth that is common across southern England and Wales, becoming more localised in the far south and west. In northern England, it is generally scarce and localised, and is absent from Scotland and Ireland.

The forewings are grey with a number of irregular zigzag lines running across them. These lines are generally black or dark brown in colour, bordered by white. This mottled appearance provides excellent camouflage on trees. However, the exposed hindwing is deep crimson in colour with broad black bands across the central section and trailing edge of the wing, and has a narrow white border. When seen from below, both the fore- and hindwings are banded black and white, and are very conspicuous.

Red Underwing caterpillars are pale brown in colour with rather wart-like nodules on the back. They feed on various willows and poplar trees.

This species can be seen in woods, parks, and by rivers, as well as in gardens. Adults have been known to feed on buddleias on sunny summer days. Like the Large Yellow Underwing, the Red Underwing flies off at great speed when disturbed.

Large Crane-fly

Tipula maxima Length: up to 30mm; wingspan: nearly 70mm
Flight period: April to August

The Large Crane-fly is the largest of over 300 species of Crane-fly – popularly known as Daddy-long-legs – found in Britain and Ireland.

Its most obvious feature are its very long, incredibly fragile legs, which are so delicate that they break off easily, even on a living insect. The body is grey-brown, long, thin and quite tapered. The narrow wings are largely clear and translucent.

Large Crane-flies are found in almost any wooded area across the country, and are frequent visitors to gardens. Adults take to the wing in early spring, while the larval forms live below ground, feeding on plant roots.

Common Crane-fly

Tipula paludosa Length: 25mm Flight period: year round

The Common Crane-fly is *the* Daddy-long-legs in many respects. Although other species pick up the name, the Common Crane-fly is the insect that truly owns the title. Both sexes have long, fragile legs and narrow wings with a brown leading edge. The male's abdomen is rather square-ended, while the female's is clearly pointed – this is her ovipositor. The wings of the female tend to be shorter than the abdomen.

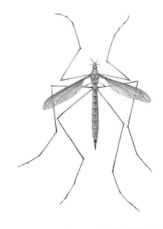

Adult Common Crane-flies are on the wing throughout the year and frequently appear inside houses, especially in summer and early autumn. They favour grassland habitats (frequently in large numbers), parkland and gardens.

The larvae live below ground, and are commonly referred to as the 'leather-jacket'. They can do enormous damage to crops as they gnaw away on the roots of their chosen crop. By adulthood, they leave their bad ways behind, and do no damage to anything whatsoever.

Common Gnat

Culex pipiens Length: 6mm Flight period: March to December

Common across Europe and seen in almost every imaginable location, the Common Gnat is perhaps one of the most irritating insects you will come across. A nondescript insect, the Common Gnat has (for its size), long legs, a chunky body, long, narrow wings and, on the male only, feathered antennae.

Common Gnats lay their eggs on water, the female producing large rafts of eggs at any one time. The larvae live beneath the water surface.

Adults rarely bite humans, much preferring to tackle birds for their meals. During the colder months of the year, they hibernate in outhouses and sheds, but they have become increasingly familiar year-round as the temperature of the planet rises.

The similarly-sized Chironomid Midge (*Chironomus plumosus*) is abundant across the whole of Europe. Resembling the Common Gnat, it is slightly larger (by around 2mm), with less prominent feathered antennae, slightly shorter legs and wings. Chironomid Midges lay their eggs in water, though the area can be small – a garden water butt is sufficient. They will often form huge dense clouds above water in summer and autumn.

St Mark's-fly

Bibio marci Length: 10–12mm Flight period: late April to May

The St Mark's-fly is perhaps one of the commonest sights on a warm, late April day. As the temperature rises, suddenly a mass of black flies appear around roadside umbellifers, all flying with legs dangling down, giving a rather lumbering appearance to the insect.

The species is entirely black bodied, except for red-hued compound eyes and translucent wings. The long black legs always dangle down in flight, giving the insect a rather lumbering appearance.

The primitive, large-headed larvae live below ground, and can cause damage to the roots of crops.

The species is so called because of its appearance on or around St Mark's day, which falls on 25th April. St Mark's-flys are common in and around gardens, as well as roadside verges, hedges and woodlands.

Cleg-fly

Haematopota pluvialis Length: up to 10mm Flight period: May to October

The Cleg-fly is seen from spring to autumn, and it is common and widespread across the country in gardens and areas of damp woodland. In northern areas, a different species of horse-fly is more common.

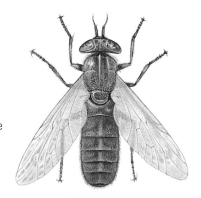

The Cleg-fly is a rather formidable insect for its size. Thick-set, and rather chunky, its greyish body contrasts with translucent wings. At rest, the wings are held above the body.

One of 28 species that fall under the 'Horse-fly' umbrella in this country, the male Cleg-fly is non-biting and feeds on plant juices and nectar. The females, however, are notorious bloodsuckers that will happily feed on an exposed arm or leg. Their flight is silent; the only warning is the dull sensation of something sticking into you. The bite itself can irritate and itch for some while after the fly has gone.

Cleg-fly larvae are as tenacious as the female. In their life underground, they prey on other invertebrates' larvae.

Window Fly and Large Bee-fly

Window Fly – *Scenopinus fenestratus* Length: 7mm Flight period: year round
(commonest from spring to autumn)
Large Bee-fly – *Bombylius major* Length: 10–12mm Flight period: April to October

The Window Fly (**below left**) is a small black fly with no bristly hairs on its body. The wings are translucent and held tightly across the body at rest. The legs are distinctly red-tinged. The species is commonly seen at the windows of houses and old buildings in particular. It is more than happy to walk rather than fly when disturbed. Its larvae are usually found in places such as a bird's nest, where they will feed readily on other invertebrates.

The Large Bee-fly (**right**) is found across much of Britain, becoming rarer the further north you go. It is certainly deserving of its name: with its stout, furry body, the Large Bee-fly bears more than a passing resemblance to a bumble bee. A closer look will rule out any bee species pretty quickly though.

The Large Bee-fly has long, almost spindly legs and a very un-bee-like probing proboscis. The body is brown and furry, and the wings, held set back from the body, are translucent except for a dark leading edge.

The species uses its long front legs to steady itself as it feeds from nectar-rich plants. Other differences from assorted bee species become apparent when the Large Bee-fly flies – note the high pitched whine and the hurried darting flight.

Large Bee-fly larvae are parasitic, feeding on wasps and Solitary Bees.

Hover-fly # 1

Melanostoma scalare Length: 6–9mm Flight period: April to November

There are almost 250 Hover-fly species in Britain and Ireland, and therefore it is impossible to cover them all within this book. The following three species are a tiny, tiny representation of the Hover-fly family, giving you a sample of the small, medium and large members of the family.

Melanostoma scalare is one of the smaller members. Both sexes have a narrow abdomen, the male's being the narrower of the two. The abdomen is black, with yellow notches on each side of the body. The thorax is black and rather bulbous, while the wings are broad and translucent.

The species is common across much of the country and is seen frequently in gardens. *Melanostoma scalare* is fond of Hawthorn blossom, and can be seen in some numbers around the tree when in flower. The larvae feed on aphids.

Hover-fly #2

Syrphus ribesii Length: 10mm Flight period: March to November

Syrphus ribesii is a common species across Europe, with many similar species occurring across the range too. It is a frequent visitor to well-flowered gardens, woodlands and hedgerows.

The rounded (almost portly-looking) abdomen is striped yellow and black, contrasting with the 'bald' thorax and reddish eyes. The wings are broad, rounded and translucent. Males habitually perch on twigs and small branches and make a high-pitched buzzing noise in flight.

Syrphus ribesii feeds on nectar and crushed pollen. The green, slug-like larvae feed on aphids, and are often themselves victim of parasitic wasps.

Hover-fly #3

Volucella zonaria Length: 15–25mm Flight period: May to November

Volucella zonaria is one of the largest Hover-fly species in Britain, but its range extends only across southern counties of England. It is a frequent visitor to gardens and flowering areas of woodland and hedgerows.

The very rotund abdomen is boldly striped yellow and black, contrasting with the 'bald' brown thorax and brown eyes. The wings are long and slightly opaque. Like many other species, *Volucella zonaria* feeds on nectar and pollen.

Celery Fly

Euleia heraclei Length: 6mm Flight period: April to November

The Celery Fly is a characteristic, if quite small, fly. The species is commonplace in gardens and open areas of countryside where umbellifers grow, and a pest to those growing celery and parsnips.

This fly has marbled, mottled brown wings, a dark brown body and green eyes. The legs are yellow-brown in colour and slightly hairy.

Celery Fly larvae feed on the leaves of umbellifers from the inside out, causing distinctive brown discolouration and holes (known by some as 'mines').

Green Parasitic Fly

Gymnochaeta viridis Length: 10–12mm Flight period: March to July

The Green Parasitic Fly is a very distinctive insect that is found in gardens, woods and parks across nearly all of Britain and Ireland.

The Green Parasitic Fly has a metallic green thorax and abdomen, which contrast with the dark brown eyes, black, bristly legs and translucent wings.

This species bears a striking resemblance to the Greenbottle (*Lucilia caesar*), which tends to be slightly smaller with a silvery hue to the eyes. Greenbottles are also more likely to be seen in rural areas than in houses.

Green Parasitic Flies are found on a wide variety of plants throughout their flight period. They lay their eggs on the plants and the larvae, when small will then burrow into, and then feed on, the caterpillars of moths.

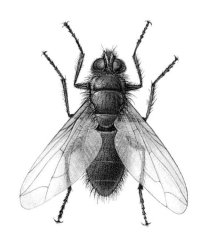

69

Common House-fly

Musca domestica Length: 8mm
Flight period: year round (commonest June to September)

The Common House-fly is one of the most familiar insects that will be seen in the house or garden. It is very common across the whole of Europe, especially where there is plenty of rotting material to feed on.

The eyes are brown toned, while the thorax and abdomen are a distinctive mix of black and tan. The wings are translucent.

Because of its feeding methods, the House-fly is a major cause of the spread of germs. It oozes saliva from a retractable proboscis. The enzymes within the saliva break up the food matter, and the resulting liquid is then sucked up by the fly. However, it often regurgitates previously digested food in the process, thereby spreading germs. The fly also defecates frequently while feeding and carries the excrement on its feet.

Bluebottle

Calliphora vomitoria Length: 12–15mm Flight period: year round

The Bluebottle is widespread across Europe, and is commonly seen in houses, gardens, woodlands and hedgerows.

Its body is deep, metallic blue-black and rather hairy. The legs are black, the eyes brownish and wings translucent.

Usually it is just females that come into houses; the males are happier outside sipping nectar. The female lays eggs on carrion, or on meat or fish left out of the fridge, which the larvae then feed on. Adult Bluebottles feed in much the same way as Common House-flies.

70

Chapter Three
Wasps, Bees and Ants

The order that covers these distinctive species is *Hymenoptera*. There are around 100,000 species within the *Hymenoptera* order and they share a number of characteristics – two pairs of wings, a mobile and comparatively large head, and toothed jaws. Some, like bees, have adapted to feed on nectar with a long tongue. *Hymenoptera* have compound eyes (as have other insects) and a set of three ocelli (simplified eyes that detect light).

There are two sub-orders within *Hymenoptera*. The first covers sawflies, such as the Horntail. The name 'sawfly' is derived from the female's long ovipositor, which she uses in the manner of a tiny saw to cut slits into plants before she lays her eggs. Many species within this sub-order are vegetarian, feeding on nectar and pollen, but others are carnivorous and feed on other insects.

The second sub-order of *Hymenoptera* contains a vast array of species, some parasitic, some solitary, some social, including bees, ants, wasps and Ichneumons. A female Ichneumon uses her long ovipositor for laying her eggs. The ovipositors of bees, wasps and ants have long been lost as an egg-laying device. They are now modified as stings, which act as a defence mechanism or as a way of stunning prey. It also includes the tiny gall wasp. This ant-like creature lays its eggs in plants and when they hatch, the tissues around the new grub begin to swell and form a gall, which in turn, acts as its larder.

Ichneumon Fly

Pimpla instigator Length: 10–25mm
Apanteles glomeratus Length: 3–4mm
Netelia testacea Length: 20mm

With over 2,000 species of Ichneumon Flies in Britain alone, it is perhaps unsurprising that they are referred to only by scientific rather than common English names. There is huge variation in size, shape and colour within the family; the tiny selection here covers those most likely to be seen in and around your garden.

Ichneumon Flies are mainly parasites of moth and butterfly larvae. This means that they are extremely important in containing and controlling numbers of insects, and they can be responsible for eliminating vast numbers of pest species. Performing this task, however, is a rather grizzly affair.

Adult Ichneumons are often seen in vegetation or on flowers, searching for unwitting hosts with their long antennae. When a female Ichneumon has found a suitable host, she raises her ovipositor, pierces the body of the host larvae and lays her eggs within it. Some larvae feed on the outside of the host, but many others feed on the inside. At first, they feed on the non-vital organs, allowing the host to survive. Then, as the Ichneumon grub nears full size, it turns its attention to the host's vital organs, killing it in the process. The grubs then pupate, often inside the shrivelled skin of the deceased host, before transforming themselves into fully winged, adult Ichneumons.

Pimpla instigator (**above left**) has a black body and contrasting orange legs. The long antennae are also black. The exposed ovipositor is around half the length of the abdomen. This species of Ichneumon is found in many habitats across Europe, and is seen from late spring to early autumn. The female *Pimpla instigator* lays as many as 150 eggs within a moth larva, the Snout Moth being a particular favourite host species.

Apanteles glomeratus (**above centre**) is found across Europe, favouring cultivated areas (including the vegetable patch). A rather small Ichneumon, the male and the female are similar in appearance, with the female generally a little larger than the male. The body is black except for a buff half-panel on the top of the abdomen. The long antennae are black, the legs pale brown and the wings translucent with a pale buff hue.

The adults emerge in two broods over the summer, and the female lays her eggs in the caterpillars of both the Large White and the Black-veined White butterflies. Up to 150 eggs are laid within the body of the unfortunate caterpillar before it is eaten from the inside out.

Netelia testacea (**above right**) is one of the commoner Ichneumon species found in Britain. It is also very conspicuous. Almost gangly in appearance, *Netelia testacea* has very long, yellowish antennae, long yellowish legs and a remarkable arched abdomen, yellow in colour with a dark brown tip.

This species is found in well-vegetated habitats across Europe and is most frequently seen on summer nights, when it is often attracted to lights in houses. *Netelia testacea* favours moth caterpillars in which to lay its eggs.

Horntail

Urceros gigas Length: female up to 40–45mm (including the ovipositor); male up to 25mm

The Horntail (or Wood Wasp) is a member of the *Siricidae* family, commonly known as sawflies. The name 'sawfly' stems from the female's long ovipositor, which has a single serrated edge used for sawing into wood before laying their eggs. The Horntail's ovipositor, however, is long and pointed, and used to bore and drill the wood instead.

Female Horntails are always larger than males. They are striking in appearance, though harmless, their body boldly patterned with yellow and black stripes, the ovipositor extending out from the rear. The legs and antennae are long, and the opaque buff wings are long and broad. The male lacks the ovipositor, and his body is almost wholly dull orange in colour, with a narrow black tip to his abdomen. His legs are black, the female's yellowish.

The female Horntail bores into ailing conifer trees to lay her eggs. Once hatched, the larvae tunnel through the timber, feeding on fungi. When they emerge, a young female will take some of the fungi with her to infect the tree that she lays her first set of eggs in.

Urceros gigas is the only native species of Wood Wasp found in Britain, though others have arrived in imported timber and become established. The Horntail favours areas of coniferous woodland but can be seen around gardens and houses, emerging from treated timbers. On the wing between May and September, the elusive males tend to remain near the treetops, while the females search for suitable egg-laying sites.

Oak Apple Gall Wasp

Biorhiza pallida Length: 1.5–3mm

This insect belongs to the family *Cynipidae*. Members of this group are usually very small, and induce gall formation on plants. A gall is an abnormal growth on a plant caused by a young insect or other organism; midges and aphids are among the other insects responsible.

The Oak Apple Gall Wasp is found, as the name suggests, exclusively on oak trees. Oak Apple Galls appear on host trees in either April or May, and several male and female wasps emerge from inside the gall during June and July. Interestingly, each gall only contains one sex, with the males appearing a day or two in advance of the females. Mating occurs, and the female then deposits her eggs within the roots of the chosen oak tree. They hatch, and form smaller galls around the roots.

Each of these root galls contains only one larva, and is only around 10mm or so in diameter. The adults that emerge in mid-winter (usually December and January) are always wingless females. These wingless insects crawl up the trunk of the tree to lay their eggs in buds, which form into Oak Apple Galls – and the cycle begins all over again.

Spangle Gall and Currant Gall

Neuroterus quercusbaccarum Length: up to 10mm

Neuroterus quercusbaccarum is another species of *Cynipidae* that is found almost solely upon oak trees. The wasp itself is comparatively large. The body and antennae are black, with contrasting brown legs. The galls are reddish brown circles, seen on the underside of oak leaves during the late summer. These are known as 'spangle galls'. Within each of the 'spangle galls' is a minute grub, which feeds busily on the gall tissue.

During the autumn, the grub and gall drop to the ground. The grub pupates and emerges as an adult (an asexual female) during the later months of winter, usually in February and March. These females lay their unfertilized eggs on oak leaves and buds (this is known as parthenogenesis). When these parthenogenetic eggs hatch, the larvae induce the formation of rounded reddish or purple galls on leaves or male catkins. These are known as the currant galls, and adult *Neuroterus quercusbaccarum* emerge from them in May and June.

The adult gall wasps from this summer generation are both male and female. They mate and, in turn, the female lays her eggs within an oak leaf. This creates a new batch of spangle galls, and the cycle starts again.

Robin's Pincushion

Diplolepis rosae Length: c.50mm

The delightfully named Robin's Pincushion is formed by a gall wasp, which is striking in appearance, with a black head and thorax and a sharply tapered golden abdomen. The legs are also golden, contrasting with the black antennae and opaque wings. Robin's Pincushion can be seen across much of Britain and Ireland, along with much of northern Europe.

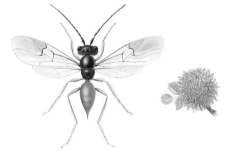

The gall itself is rather small, reddish-orange in colour, spherical in shape and hairy. Unlike the Spangle Gall and Currant Gall (*Neuroterus quercus-baccarum*), this species does not have alternating generations, and almost always reproduces through parthenogenesis. This is partly because the male *Diplolepis rosae* is rare.

The wasp's flight period is between early spring and mid-summer, with the galls maturing in the autumn. They can be seen in gardens (on roses), open areas of countryside, woodland fringe and parkland.

Black Garden Ant

Lasius niger Length: worker c.5mm; male and queens up to 12mm

The Black Garden Ant is a familiar sight, very common in gardens. The 'worker' is the ant that we see: it is black or very dark brown with a large head and abdomen and comparatively long antennae. Its narrow waist (known as the pedicel) is in one single segment. This is a non-stinging species, but it can give you a nip.

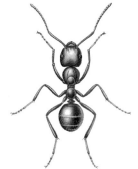

In late summer (usually July or August, depending on weather conditions) fully winged, newly emergent male and female Black Garden Ants take to the air. Often these flights of 'flying ants' occur across a large area over a couple of days, causing a mini-plague.

The flying ants are around twice the size of workers and their flights are solely in the name of reproduction. Both sexes fly off on their wonderfully named 'marriage flights' and often mate on the wing, or on the ground. For the male, life is short lived; soon after mating, he dies. The female lasts longer – she sheds her wings and looks for somewhere suitable to nest. Unfortunately, the flying ants are the favoured snack of several aerial feeding species, including Black-headed Gulls, which swarm and circle excitedly above the emergent ants' nests. The toll is often so heavy that very few females actually survive to form new colonies.

As with so many other species of ant, the Black Garden Ant is found in large colonies. A single colony can consist of many thousands of worker ants and a solitary queen. The ant is omnivorous, but particularly fond of anything sweet, and frequently 'milks' aphid species for their honeydew. The species is found across Britain and Ireland, and is widespread across Europe.

Red Ant

Myrmica rubra Length: worker c.4mm; male and queens up to 10mm

The Red Ant is another familiar species seen in Britain – one of 42 ant species seen here. A worker Red Ant is marginally smaller than the Black Garden Ant.

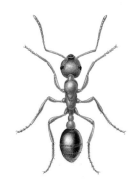

Red Ants are a rich reddish-brown in colour, except for the black, rather bulbous abdomen, which is more pronounced on the female. The male's abdomen, however, is longer. The narrow waist (the pedicel) between the thorax and the abdomen is made up of two segments, unlike the single segment of the Black Garden Ant. The Red Ant has a rather nasty nip, injecting stinging formic acid into its victims and causing mild irritation in humans.

Males and queens emerge in late summer or early autumn and are noticeably larger than the workers, approaching nearly 10mm in length. There are several queens within a colony, along with a few hundred workers. They are omnivorous, but compared to the sweet-toothed Black Garden Ant, they prefer more savoury items, such as animal food.

Red Ants are found in almost any open habitat across the whole of Europe, and are a common sight in gardens.

Common Wasp

Vespula vulgaris Length: worker c.10mm; male and queens up to 15mm

The Common Wasp is undoubtedly the scourge of summertime. Unpopular with so many of us, it is, in fact, a misunderstood opportunist. Common Wasps are found widely across the whole of Europe and generally take to the air from late spring to early autumn.

There are seven species of wasp native to Britain, although others have become established in recent years. Face patterns are often the only clue to identify wasps: head-on, the Common Wasp shows much black across the middle. Thorax markings also offer clues – note four yellow spots near the rear.

As the summer draws to a close, 'workers' are rearing male and females in special cells within their amazing paper nests. As the wasps reach maturity, the colony begins to break up. The workers have no more larvae to occupy their time, so they can turn their attention to all the sweet things they relish. This is when they can become problematic. Indeed, wasps may nest very close to you, but for much of the summer you may hardly notice their presence. As colder weather sets in, almost all Common Wasps die, leaving just the larger, mated queens to survive.

German Wasp

Vespula germanica Length: worker c.11–12mm; male and queens up to 20mm

The German Wasp is increasingly common in Britain and found widely throughout Europe, with the exception of northern Scandinavia. It resembles to the Common Wasp and distinguishing the two can be tricky.

Generally, German Wasps are slightly bigger than Common Wasps, with the face of the German Wasp showing less black – this is perhaps the best way of telling them apart.

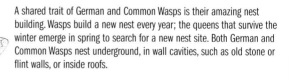

A shared trait of German and Common Wasps is their amazing nest building. Wasps build a new nest every year; the queens that survive the winter emerge in spring to search for a new nest site. Both German and Common Wasps nest underground, in wall cavities, such as old stone or flint walls, or inside roofs.

Once the queen has chosen her site, she gnaws and scrapes away at fence posts or trees, chewing the wood and mixing it with saliva to form a pulp, which she spreads out in a thin paper layer. After forming an initial dome, she adds cells, laying an egg in each one. She then feeds the larvae, which become the workers; they take over the building of the nest, allowing her to go into egg-laying overdrive. An extraordinarily beautiful football-sized structure, the nest is encased by a thin layer of paper.

Estimates suggest that up to 20,000 wasps can emerge from one nest during a season, but not all at once. Young wasps are reared on a mass of insects, many of which may be harmful to a garden – proving the wasp is more friend than foe.

Hornet

Vespa crabro Length: worker up to 25mm; queens up to 50mm

The Hornet is an elusive but unmistakable member of the Wasp family, because of both its size and colouration. Luckily, Hornets are far more placid than either the Common or German Wasp.

Hornets share the yellow abdomen of the smaller wasps, but the colour is a little more subdued. The head, thorax and stripes on the abdomen are a rich chestnut brown in colour, as are the legs and antennae.

Unlike its smaller cousins, the Hornet prefers to build its nests in hollow trees, but will also choose chimney breasts, roof spaces or wall cavities. Adults prey on insects as large as butterflies and dragonflies to feed their young.

Hornets are found across Britain and much of Europe, although they are absent from Scotland and Ireland, as well as northern Scandinavia. This wonderful insect can appear in gardens at the edge of woodland, in larger, planted gardens or in woods and parkland.

Tawny Mining Bee

Andrena fulva Length: female c.10mm (slightly larger than the male)

The Tawny Mining Bee is a rather small member of the bee superfamily *Apoidea*. There are at least 30,000 different species in this superfamily around the globe, with around 253 of these found in Britain and Ireland. The Mining Bee genus, *Andrena*, accounts for around 60 species in this country.

Unlike the Honey Bee, Tawny Mining Bees are solitary. The head is black, contrasting with the dark orange thorax and abdomen. The legs are blackish and the wings are translucent. A male Tawny Mining Bee is smaller and less colourful than the female, lacking all the tawny tones – his body is blackish, bordered by tawny hairs.

Tawny Mining Bees are found in almost any open habitat and are common in central and southern Europe. They are found across much of southern and eastern England.

The flight period for this little bee is from April to late June or early July. They nest in the ground, favouring light, sandy soils. They frequently nest in lawns – look out for mini volcano-like mounds of excavated soil; although there may be many of these mounds across the lawn, they do not represent communal life. Each of the females works alone, excavating half a dozen chambers below ground. She leaves her progeny in a cell full of food, and shows no interest in their welfare thereafter; indeed, she is likely to have died before they emerge.

Honey Bee

Apis mellifera Length: queens c.20mm; drones c.15mm, workers up to 10mm

The Honey Bee is perhaps one of the most studied insects in the world. Countless authors have written books and articles on its behaviour, social interaction and the incredible methods of communication the species uses to locate the next good nectar source. Found across the whole of the European continent, it is thought that the Honey Bee actually originated somewhere in rather warmer climes, because of the perennial nature of its colonies.

There are three types of Honey Bee within the hierarchy of a colony. At the top is the queen. The drones (males) support her, and beneath them come the workers. These three vary widely in appearance. Queens are generally about 20mm in length, with a dark body and paler legs. The drones have the 'classic' look of the Honey Bee, with a neatly striped black and tan body. The smallest of the three, the worker bee is like a smaller version of the queen, with a dark, almost black, looking body.

Honey Bees are on the wing from early spring to late autumn, depending on weather conditions. In their truly wild state (i.e. not in a hive) Honey Bees create their nest in the hollow of a tree, and a well-established colony is thought to contain as many as 50,000 bees.

The cells within the nest are perfect hexagons, and are used for rearing young or storing honey. Other cells exist within the nest too, for the drones and also for the queen. Eggs laid within the queen cells have their future dictated by the worker bees: some of these eggs will be chosen by the workers to become queens themselves, in which case the larvae are fed a nourishing mix of protein and fatty secretions by the attendants (known as royal jelly).

The queen will always either come into a colony or takes parties of workers with her to start a new colony elsewhere. The focus of the colony's attention, the queen, does little to justify it. She can live for many years, but drones and workers are short lived, especially during the hectic summer months. However, despite her rather relaxed way of life, the queen does not get everything her own way. If there is a hint that she is failing, the workers rise up, kill her off, then welcome a new queen to the colony.

Garden Bumble Bee

Bombus hortorum Length: c.20–25mm

The Garden Bumble Bee is one of several species that belong to the family group *Bombus*. Most species share the familiar black and yellow horizontal stripes on the body, though this is not the case with all the species within the group. This species is widely encountered in gardens and well-vegetated areas of grassland across Europe.

The thorax has two broad yellow stripes separated by a black one. The abdomen has stripes of three different colours – yellow closest to the thorax, black in the middle of the abdomen and white at the rear end. The legs and antennae are dark, the wings translucent.

The Garden Bumble Bee, like many other species, nests underground. In spring, the female emerges from her winter hibernation and sets about finding a new nest site for the summer ahead. She will collect grass and moss to build her nest, which she will construct in a crevice or an old mouse hole. The queen then laces the small ball of nest material with pollen before she lays around 10 eggs. Once the eggs are laid, the queen will surround them with a waxy substance she produces from glands on her abdomen. After less than a week, the eggs will hatch (following a period of incubation from the queen) and the larvae will grow quickly on the pollen and nectar stored for them. Within in a fortnight they have pupated and are ready to emerge.

As the queen continues to lay more eggs, her worker offspring take over the domestic chore of feeding the new larvae; as the season progresses, the workers become bigger and bigger with each brood, and at the end of summer, new queens emerge. After taking to the wing and mating with recently emerged males, a new queen takes herself away to winter quarters, and the cycle begins again.

79

Buff-tailed Bumble Bee

Bombus terrestris Length: c.20–25mm

The Buff-tailed Bumble Bee is similar in size to the Garden Bumble Bee, but the markings are quite different. The mainly black thorax has a more orangey 'collar'. The first segment of the abdomen is black, and the broad second segment is orange-yellow. This is bordered by black on segment three; the 'tail' is buff and particularly noticeable on the female. On the continent, rather confusingly, the female has a white tail tip.

Buff-tailed Bumble Bees are found in almost any well-vegetated areas across the whole of Europe.

Queens emerge as early as February or March, when they visit the catkins on sallow. The workers that follow feed on apple and cherry blossom.

Like the other species of Bumble Bee detailed here, the Buff-tailed Bumble Bee nests below ground.

White-tailed Bumble Bee

Bombus lucorum Length: c.20mm

The White-tailed Bumble Bee is similar in size to both the Garden Bumble Bee and the Buff-tailed Bumble Bee. It is very similar in appearance to the Buff-tailed Bumble Bee and it is really just the colour of the 'tail' that separates the two – the white of the White-tailed Bumble Bee is very white indeed, and is very conspicuous in comparison with the buff of the previous species.

White-tailed Bumble Bees are seen in well-vegetated areas across Europe.

The queen emerges as early as February, and, like its Buff-tailed cousin, feeds on sallow catkins.

This species also nests below ground.

Chapter Four
Beetles

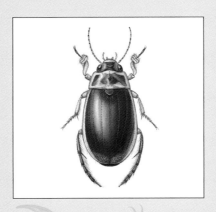

There are over 350,00 species of beetle worldwide, with nearly 20,000 found in Europe, and around 4,000 in Britain and Ireland. In fact, *Coleoptera*, the order of insects that beetles belong to, has the world's greatest number of species.

Most species of beetle can fly, but don't spend much time doing so. Many of them are found on the ground, where they live in vegetation, leaf litter, or beneath logs or stones. Flight is used primarily as a way of finding food or a mate.

Like other insect species, beetles have compound eyes and antennae. The size and shape of these antennae are useful when trying to identify some of the trickier species.

The legs of different beetles can give clues to the behaviour of that species. For instance, beetles that are able to swim have flat, almost paddle-like legs, while beetles that need to dig have toothed front legs to aid their progress.

Several species are widely regarded as pests – Cockchafers damage crops, several species infest grain and flour mills, while Carpet Beetles can damage many different fabrics, not just carpets. Perhaps the worst pest of all is the Woodworm, where the developing larvae are able to do untold damage as they reach adulthood. However, it is not all doom and gloom. Ladybirds, for instance, are a gardener's friends, munching their way through other insect pests such as Greenfly.

Black Beetle

Feronia nigrita Length: c.15mm

Black Beetles are found across Europe, in habitats ranging from gardens to woodland and parkland. They are nocturnal creatures and feed almost exclusively on other invertebrates.

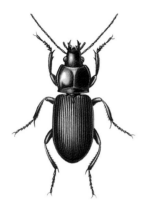

The Black Beetle is a rather distinctive beetle, with an entirely black body and legs. To the inexperienced eye, it bears more than a passing resemblance to the larger Carabid Beetle, but differs in size, overall colour and, perhaps, most distinctive of all, the neat grooves that line the main 'shell' of the bug - known as the elytra. The elytra (singular elytron) are actually beetles' tough forewings.

Interestingly, the Common Cockroach *Blatta orientalis* is occasionally referred to as the 'Black Beetle', but *Feronia nigrita* is the rightful owner of the name.

Devil's Coach Horse

Staphylinus olens Length: 25–30mm

The Devil's Coach Horse is one of the most striking and evocative names of any beetle found in Britain, and indeed, Europe. This insect is frequently seen in gardens, hedgerows, parks and woodland.

One of nearly 1,000 species of beetle found in Britain that belong to the *Staphylinidae* family (there are over 2,000 across continental Europe), the Devil's Coach Horse is the largest member of the family; the smallest measures in at just under 1mm in length. It is completely black in colour, and its body appears to be in four main parts. Particularly noticeable are the small, short elytra and the long abdomen.

The Devil's Coach Horse also has a distinctive alarm posture - when under threat, it arches its tail over its back and simultaneously opens its jaws.

As with the Black Beetle, this species is a natural born predator, and will feed on other invertebrates and slugs.

Carabid Beetle

Carabus nemoralis Length: 20–30mm

Comprising over 350 different species in Britain alone, the Carabid Beetle and other members of the family *Carabidae* are often known as ground beetles, as they spend so much of their time on, or sometimes below, ground. The Carabid Beetle is found in most European countries, with the exception of northern Scandinavia, and in a wide variety of habitats.

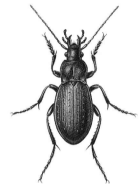

The Carabid Beetle (a member of the *Carabidae* family) resembles the smaller, blacker Black Beetle, with a blackish looking body, frequently tinged with a metallic purple, bronze or green sheen. Unlike the grooved elytra of the Black Beetle, those of a Carabid Beetle are finely ridged and have tiny ornamental raised 'dots'. Female Carabid Beetles are less shiny than males.

This very quick, flightless beetle is a nocturnal predator, like many members of its family.

Common Cockchafer

Melolontha melolontha Length: 20–30mm Flight period: May
(hence their alternative name of Maybug) to late July (sometimes into August)

Common Cockchafers are widespread across the whole of Europe, but only to around 1,000m – the cold countries of northern Scandinavia are not to their liking, and they are absent from the region. Gardens are a favourite haunt of Common Cockchafers, but they also enjoy parkland and woodland.

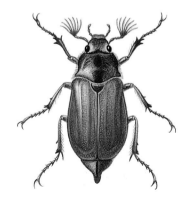

The Common Cockchafer is well known to those who leave windows open in summertime. The crash as the beetle hits the lampshade, coupled with hefty size and loud, buzzing flight make it an imposing creature. It is colourful too, the black thorax contrasting with the rust-coloured elytra. It has brown legs and fanned antennae, which are larger on the male than on the female.

Notoriously, adults have a voracious appetite for leaves and flowers and cause much damage to trees and plants. The big, fat, white larvae are even worse; as they develop to adulthood, the larvae spend between three and four years underground, happily eating the roots of cereal crops and grasses.

83

Stag Beetle

Lucanus cervus Length: 20–30mm or even larger (up to 80mm is possible)
Flight period May to August

Stag Beetles breed in rotting tree stumps but, sadly, seem to be declining quickly across their range, which extends from England through central and southern Europe.

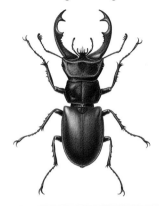

The Stag Beetle is perhaps the most exciting and fearsome-looking beetle found in the garden. With its large size, black head and thorax contrasting with deep woody brown elytra, it is an impressive sight. Add to this the amazing antler-like jaws (hence the name Stag Beetle) and you have a very special creature indeed.

The large antler-like jaws of the male give it a rather frightening appearance, but the 'antlers' are often used just for show, although they are used to impress females, or for wrestling other males. They cannot be closed with any real force. However, the female's smaller antlers can give you a nasty nip.

Oak woodlands are their ideal habitat, but they are also seen in parks and gardens. Adults feed on tree sap; the fat, white larvae feed on rotting wood.

Lesser Stag Beetle

Dorcus parallelopipedus Length: 20–30mm Flight period: April to October

The Lesser Stag Beetle and the Stag Beetle are two of three members of the *Lucanidae* family to be found in Britain. The third is *Sinodendron cylindricum*, which resembles a cross between the two species. Lesser Stag Beetles take to the wing for a slightly longer period than the larger Stag Beetle, and can be seen in gardens, woodland and parkland across central and northern Europe.

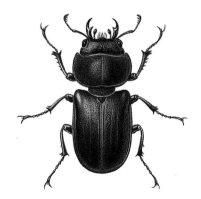

As the name suggests, the Lesser Stag Beetle is a slightly smaller version of its cousin. While it can match the Stag Beetle in size, it tends to be smaller. As well as lacking the prominent 'antlers', the Lesser Stag Beetle also lacks the rich brown elytra of its relative, and has fewer of the distinctive 'spurs' on the tibia of its legs. It is entirely black in colour, with the male having a large, wide-headed look.

Adults feed on tree sap and the brown-headed, white-bodied larvae live in, and feed on, rotting wood.

Summer Chafer

Amphimallon solstitialis Length: 15–20mm Flight period: May to August

Found across much of Britain, Ireland and Europe, the Summer Chafer is common in gardens, as well as areas of woodland, parkland, hedgerows and scrub.

The Summer Chafer is similar in many of its habitats and behaviour to the more colourful Common Cockchafer, though it is smaller. A rather hairy looking insect, the Summer Chafer is tawny brown all over (lacking the colour contrasts of the Common Cockchafer), and its smaller antennae appear much less fanned.

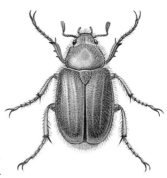

Just like the Common Cockchafer, adults are attracted to light, and will merrily buzz around any room in the house where the lights are on and the windows open. Summer Chafers will often swarm around trees and bushes at dusk, and throughout the hours of darkness.

Summer Chafers take up to two years to mature from larval stage to adulthood and, like the Common Cockchafer, survive by eating roots.

Two-spot Ladybird

Adalia bipunctata Length: 3.5–5mm Flight period: early spring to late autumn

The Two-spot Ladybird is one of Europe's commonest ladybirds, and is found at almost every compass point across the continent.

The species is quite small and rather variable, the type most frequently seen having single bold black spots on each of its pillar-box red elytra. The head is black with bold white patches on the side. Note, too, the black leg colour, a useful clue to separating this from other ladybirds. In more northern populations, the ladybirds become darker and darker, with some appearing wholly black.

Two-spot Ladybirds are found in almost any well-vegetated habitat and feed on aphids that favour herbaceous plants. During the winter months, a garden shed or an outhouse can house massive groups (usually many hundreds) of ladybirds in semi-hibernation mode. The larvae are small, blue-grey in colour with pale cream to yellow spots.

Twenty-two-spot Ladybird

Psyllobora vigintiduopunctata (more commonly Psyllobora 22-punctata)
Length: no more than 3–4mm Flight period: April to early September

The Twenty-two-spot Ladybird, like several other ladybird species, has a scientific name that boasts numerals as well as digits. A tiny insect, the Twenty-two-spot Ladybird has a bright yellow body with 11 (sometimes 10) black spots on each elytron. The pronotum (the first thoracic segment) is also yellow with five black spots.

Sharing similar vegetated habitats to other species of ladybird, the Twenty-two-spot Ladybird is found across Britain and Europe.

It spends the winter tucked away in leaf litter, though it may be seen in milder winters. This particular species feeds almost entirely on the mildews found on umbellifers and low growing shrubs.

Seven-spot Ladybird

Coccinella 7-punctata Length: 5–8mm Flight period: early spring
(usually in April, but earlier if conditions are good) to September or October

The Seven-spot Ladybird is one of the largest ladybirds found across Britain and Europe and is certainly one of the commonest.

The species is similar in appearance to the much smaller Two-spot Ladybird. The elytra are bright red and each have three bold black spots, with a larger black spot spanning the two forewing cases. The head is black, with less white on the sides than the Two-spot Ladybird. Like the Two-spot Ladybird, the Seven-spot Ladybird has black legs.

Seven-spot Ladybirds are found in well-vegetated areas, often in considerable numbers. At the height of some of Britain's hottest summers (most recently 2003 and 2004) massive invasions of many millions of ladybirds (mainly Seven-spots) occurred, even making headlines across the media.

Adults and larvae alike feed on aphids (Greenfly being a real favourite) and this makes them especially popular with gardeners. The larva is small, blue-grey in colour with pale cream to yellow spots. Seven-spot Ladybirds spend the winter alone or in small groups tucked away in leaf litter, or inside a shed or a house.

Wasp Beetle

Clytus arietis Length: 5–15mm Flight period: May to August

Found across Europe (except for northern Scandinavia), the Wasp Beetle is a member of a group known commonly as 'longhorn beetles', which boasts around 60 species in Britain and Ireland.

Amongst these species, the Wasp Beetle is unique in its uncanny, mimicked wasp-like appearance, but unlike the Common Wasp, it is harmless.

Wasp Beetles have a long central body and long legs. The head is black, and the yellow, wasp-like patterning on the black elytra is variable; some individuals look almost black, while others are boldly patterned. The Wasp Beetle's habit of scuttling over tree trunks and leaf litter further enhances its mimetic resemblance to a wasp.

Found in well-vegetated gardens, woodland and hedgerows, the Wasp Beetle feeds on nectar and pollen. Females lay eggs in dead wood or, sometimes, living trees to which the burrowing larvae can cause considerable damage.

Furniture Beetle

Anobium punctatum Length: no more than 5mm Flight period: late April to August

The Furniture Beetle may be tiny and look rather insignificant, but it is one of the most notorious beetles for creating havoc around the house. It is the Furniture Beetle's larvae that cause all the problems; these larvae are the notorious woodworm.

Adult Furniture Beetles are variable in colour, ranging from dark brown to straw yellow. The elytra are finely ridged and the longish (for the size of the beetle) antennae are clubbed in shape and covered in very fine hairs.

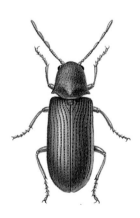

Adults are seen from spring to late summer across central and northern areas of Europe, and their favourite habitat is the dry wood of coniferous and deciduous trees alike. They are, of course, also abundant in houses.

The beetle lays its eggs in any small crevice within a branch, a cupboard or a rafter. The larvae, once hatched, tunnel through the dead wood and turn whatever they may be residing in to sawdust. It is not until the newly fledged adults leave their destructive worm-holes that they make their presence known. By then it is often too late...

Great Diving Beetle

Dytiscus marginalis Length: 25–35mm

The Great Diving Beetle is a magnificent addition to any garden pond, even if they are somewhat tenacious and voracious towards other pond inhabitants. It is Europe's commonest diving beetle (up to 150 species are found across the continent, 110 in Britain alone) and are found in well-reeded ponds or larger areas of still water.

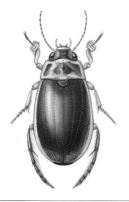

Great Diving Beetles vary from very dark brown (with an olive hue) to blackish in colour, with a distinctive pale cream border to the thorax and elytra. The male is larger than the female and has smooth-looking elytra, whereas the female's have a distinctive ridged appearance.

The larvae of the Great Diving Beetle look rather like dragonfly larvae, but their segmented bodies give them better mobility. They live underwater and, like the adults, are fearsome predators. Both adults and larvae prey on tadpoles, other invertebrates, and even newts and fish. However, their methods of feeding are quite different. Adults attack with their powerful jaws and devour their prey through the mouth, while the larvae use the gruesome tactic of closing their sharp mandibles into their prey, 'injecting' it with digestive juices and sucking it dry.

Whirligig Beetle

Gyrinus natator Length: just over 6mm

Whirligig Beetle is actually a generic term for the dozen or so species of the group *Gyrinidae* that are found in Britain, but it is *Gyrinus natator* that is most frequently referred to - a tiny, instantly recognizable garden pond beetle.

Shiny black all over, adults spend the majority of their lives spinning endlessly in circles on the water's surface. If you manage to see one at very close quarters, note the wonderfully well-adapted middle and rear legs - both sets are small, hairy and paddle shaped, enabling terrific motion across a pond. Another classic design feature is the beetle's ability to see above and below water simultaneously, as each eye is divided into two parts.

Whirligig Beetles feed primarily on small insects that fall onto the pond surface, but they are also capable of diving for prey. Diving is also used as a defence mechanism. This particular beetle can be seen throughout much of the year, often in large numbers as summer turns to autumn. In the depths of winter, adults burrow into the muddy bottom of the pond to hibernate, emerging to lay their eggs on pond vegetation in the spring.

Chapter Five
Spiders

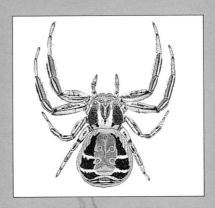

Spiders are much-maligned creatures, the cause of one of the most frequently occurring phobias known to the medical world, and, all too often, unjust victims.

The spiders of northern Europe are completely harmless. Along with harvestmen, scorpions, ticks and pseudoscorpions, they come within the class *Arachnida*. Members of the *Arachnida* class share some features with insects, particularly the tough exoskeletons made of chitin. The most obvious differences are usually easy to see. All members of the *Arachnida* class have four pairs of legs rather than three, and the bodies are in two parts rather than three.

As well as having eight legs, *Arachnida* have eight eyes (though some species may have only six) but these eyes are not as sophisticated as insect eyes. Spiders and their like have an excellent sense of touch, and they are able to detect the subtlest of vibrations.

Spiders lay their eggs in a cocoon or silk sac. These eggs may be dormant throughout the winter months and some females guard them religiously. She will also take care of her young for some time after they have emerged from their cocoon, and they may eventually disperse on air currents, by moving to the end of a branch and creating their own silk strand before being caught in a current of air and drifting away.

Green Orb-weaver

Araniella cucurbitina Length: 3.5–3.6mm

The Green Orb-weaver (also known as the Orb-web Spider) is a distinctive spider – rather small, but very colourful. The species can be found in any low vegetation, bushes or trees across the whole of Europe.

Female Green Orb-weavers are just a little larger than males, with a larger abdomen, and whereas the male has an orangey head, hers is browner. Both sexes share four pairs of brownish-grey, slightly hairy, legs. The abdomen is basically lime green with a paler green stripe on each side. The paler stripes are marked with a few dark blobs, and in the centre of the abdomen are up to three or four dark blobs surrounding a distinctive inverted black 'Y' shape.

The adults are seen from early summer to late autumn, when they weave their small and rather untidy orb-shaped web. Egg sacs are laid on the underside of fresh leaves and then encased with layer upon layer of silk. Eventually the leaves fall to the ground and the young spiders emerge in spring. As with many other spider species, the female dies once she has laid her eggs.

Garden Spider

Araneus diadematus Length: male up to 5mm, female 10–15mm

The Garden Spider is one of the most distinctive and beautifully marked spiders seen around the garden (or anywhere else come to that). It is commonly found in gardens, hedgerows, woodland and heathland across northern Europe.

Garden Spiders are one of the larger spiders covered in this book, but it is by no means the biggest. As with several other species, the female Garden Spider is larger than the male. As with the Green Orb-weaver, the male Garden Spider has a smaller abdomen than the female. The Garden Spider varies in colour from straw yellow to dark woody brown. The legs are stripy and the head is quite pale. The abdomen on a 'standard' tan and brown Garden Spider has a beautifully intricate and complex pattern of tan, brown, black and white, with the white markings resembling a cross.

They prey on insects such as aphids and flies, catching them in their delicate webs. During the autumn months, a female lays many hundreds of eggs (up to about 800) in one mass and then protects them with a silk binding. After remaining with her eggs for a month or so, the female Garden Spider then dies.

Missing Sector Orb-weaver

Zygiella x-notata Length: male just under 5mm, female c.9mm

The wonderfully named Missing Sector Orb-weaver is, at first glance, a rather unremarkable spider, but a closer look reveals a number of subtle, intricate and beautiful markings. The species is found across much of Europe with the exception of some of the continent's more northerly nations.

One of the most noticeable features of this spider is its very long front legs. Of all the species covered in this book, it is the only one with a combination of three short sets of rear legs and one long set of front legs. The head tends to be pale brown in colour, while the thorax has a dark brown central wedge bordered by pale tawny brown. On the abdomen there is a broad greyish central streak, bordered by black. The upper-most sides of the abdomen are decorated with a number of thin, greyish 'C' shapes. As with the two species mentioned above, the male is smaller than the female.

The Missing Sector Orb-weaver is very fond of human habitation as well as gardens, parks, hedges and woods. This fondness for human dwellings accounts for many of the webs that you see hanging in the corner of a window frame or across the top of a door. The species' common name comes from the fact that it deliberately leaves missing sectors at the top of its webs in the hope that prey will stumble into the trap.

Crab Spider

Xysticus cristatus Length: male 3–4mm, female 6–8mm

As its name suggests, the Crab Spider is a small *arachnid* that is somewhat crab-like in appearance. The Crab Spider is widespread across Britain, Ireland and northern Europe, favouring low bushes and plants, as well as being more than happy on the ground.

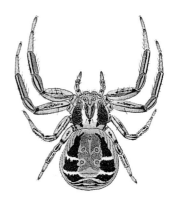

Both sexes can have rather variable colouration and markings, but tend to be pale brown with black and white abdominal marks. The legs are pale brown, and the two front pairs are held a little, like those of a crab – angular, and poised to catch prey at any time. The head and thorax are generally brown with paler internal marks, while the abdomen has a pale brown central mark edged by broad black and thin white stripes.

Seen during spring and summer, a Crab Spider will take cover and hide on flowers before jumping and pouncing on tiny aphid and insect prey.

Nursery-web Spider

Pisaura mirabilis Body length: male 10–20mm, female at least 20mm

The Nursery-web Spider is a rather large spider and is often only just smaller than the House Spider. This species is found across northern Europe and favours many different habitats, including woods, parks, heaths and grassland as well as the garden.

Nursery-web Spiders range in colour from pale straw to dark brown. Their abdominal markings can also vary widely, from clear and distinct to virtually non-existent. Generally, the markings include a vertical white line on the thorax, with a complex pattern of browns and cream on the abdomen, the cream bordering the dark brown central panel. This central panel frequently shows buff internal markings. Like the House Spider, both sexes have very long front and back legs with slightly shorter ones in between. The male is smaller than the female and has a narrower abdomen.

Seen throughout the summer months, the Nursery-web Spider is a daytime hunter, running swiftly for prey; it also likes to sun itself for long periods on trees or plants. The female carries her eggs in a cocoon that she hangs from her fangs before spinning a silk shroud over it. She then remains with the cocoon, on guard, until the young have hatched and moved away.

House Spider

Tegenaria duellica Body length: at least 20mm

The House Spider (or Cobweb Spider) is undoubtedly the monster of the pack! This is the spider that you will see scuttling across the living room floor in the evenings, particularly in the autumn and early winter months, when males are on the prowl for a female. Very common and widespread across northern Europe, like the Missing Sector Orb-weaver, it is found in houses as well as gardens and woodland.

House Spiders are rather uniform in colour. They have very long, sandy brown front and back legs, with slightly shorter, similarly coloured central pairs. The head is brown, with a slightly darker set of markings on the thorax, while the abdomen is blackish-brown in colour with a pale central streak and is barred on either side.

House Spiders feed by trapping their prey in a triangular tunnel-like web with a little hideaway for the spider at the base, where it waits until its prey becomes entangled in the web. The web is another familiar sight around the house. Unlike the species mentioned previously, House Spiders are long-lived creatures; a female can live for many years.

Harvestman

Phalangium opilio Length: male as little as 3mm, female 10mm or more

The Harvestman is a small, extremely long and thin-legged arachnid. It can be found across Europe, and thrives in long, dense vegetation.

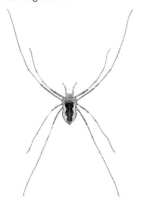

The body of a Harvestman is tiny, with grey to yellow sides and a black zigzagging central stripe. The underside of the body is white. The long, incredibly thin legs are its most striking feature. Typically, the female is larger than the male, which can be as small as a male Crab Spider.

Feeding during the hours of darkness, a Harvestman preys on other small invertebrates. This species will winter in egg form, maturing during the months of the following summer.

Daddy-long-legs Spider

Pholcus phalangioides Length: up to 15mm or more

The Daddy-long-legs Spider is similar to the Harvestman, but is on average a little larger with a wholly pale abdomen and even longer, thinner legs. Present across central and southern Europe, this species is frequently found in houses where they are very much at home living in corners.

Daddy-long-legs Spiders are in fact entirely pallid in tone; their cylindrical abdomen, head and legs are all pale greyish-yellow.

Both sexes catch prey by hanging upside down from a thin web, from which they catch other small spiders and insects. The spider then entwines its victim in spun silk. The male is only seen during the spring and summer, the female throughout the year.

Chapter Six
Amphibians and Reptiles

Amphibians are the most primitive land-living vertebrates. They are divided into two classes – the *Urodela*, which includes newts and salamanders, and the *Anura*, to which frogs and toads belong.

Newts are long-bodied creatures with soft, scaleless skin and strong tails. They breed in water and lay eggs on stones or aquatic plants. While frogs and toads also breed in water, they are rather different in appearance to newts. The most immediate difference is their stockier body shape; they also lack tails and can spring and jump.

Reptiles, for the most part, have adapted to a terrestrial lifestyle. They can be immediately distinguished from amphibians by their scaly appearance. Reptiles also have a markedly different reproductive cycle. In many species, the young hatch from eggs (which are fertilized internally), although the Viviparous Lizard gives birth to live young.

Lizards have two pairs of legs (although they may not be visible, as in the case of the Slow-worm), while snakes are legless. Neither group can manage their body temperature internally, so to maintain a constant core temperature they bask in sunlight. In northern areas, amphibians and reptiles hibernate, as survival is almost impossible in very cold climates. Most choose to hibernate in a simple hole in the ground or a crevice in a wall. Frogs may hibernate underwater, at the bottom of a muddy pond or a ditch, while toads have been found in old birds' nests.

Common Newt

Triturus vulgaris Length: c.7–11cm

Common Newts are found from Britain and Ireland to central areas of Europe, but they are absent from southern Europe and northern Scandinavia. They are fairly terrestrial and can be found on land in many different damp places, from gardens and woods to fields and rocky areas. Common Newts feed on worms, fresh water molluscs and crustacea, caterpillars and insects.

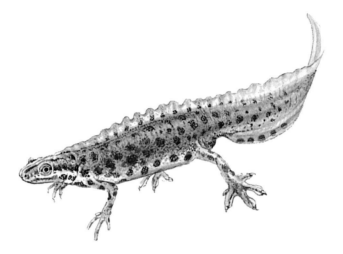

The Common Newt (also known as the Smooth Newt, due to the smooth, velvety feel of its skin) can vary considerably in both size and appearance, with distinctively different markings. This is especially the case in more southern and eastern countries within its range; in Britain and northern Europe it is more consistent in size and appearance. Males are larger than females, and are more strikingly patterned, especially in the breeding season, which begins in early spring.

A breeding male Common Newt has many bold black spots along the length of his body, except for the very tip of his tail. The wavy crest on his back is also boldly spotted, the black contrasting with the olive tones of the crest and flanks. The lower flank is paler, fading to creamy white on the belly. In high season, the male Common Newt will show a broad orange stripe down the centre of his belly. There are black spots on the belly too, but these vary from big and bold to small and insignificant. The olive legs and toes are also spotted. A breeding female will be a rather drab affair in comparison to the male. She lacks a crest and any of the male's bold black spotting, and is pale olive brown to yellowish brown in tone. She does, however, have some colour on her underbelly in high season. Outside of the springtime breeding season, both sexes look less impressive. The male's crest is greatly reduced and both sexes are less boldly patterned and marked.

Common Newts are long lived and can survive for 20 years or so, becoming sexually active from their third year onwards. After elaborate displays in still water ponds (the weedier the better), the male releases his sperm, which the female collects. The female lays individual eggs on the leaves of aquatic plants. After around 21-28 days, tadpoles emerge and, within 10 weeks, they have grown and transformed into newts. This cycle normally ends in September, and by October the time has come for hibernation.

Great Crested Newt

Triturus cristatus Length: 14cm, some females recorded at 18cm+

The Great Crested Newt (sometimes known as the Warty Newt) is a large and impressive-looking amphibian that is found across Europe, with the exception of Ireland, northern Scandinavia and Iberia. In Britain this species was formerly critically endangered but thankfully, after many years of proactive conservation projects, its future looks much brighter.

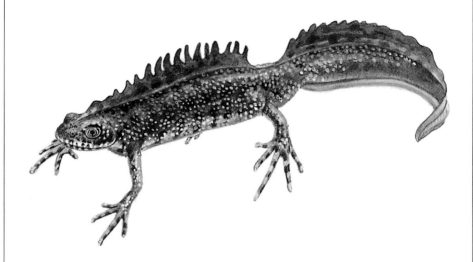

Compared to the Common Newt, the skin of a Great Crested Newt is rather coarse and, as suggested by its alternative name, warty too.

Large and dark, Great Crested Newts are typically dark brown above (looking black when seen on dry land) with contrasting bright orange underbellies. Both the sexes have black spotting on their upper bodies, with bold black spots on the orange underbelly. In the breeding season, the male's crest develops in all its ragged glory along his back and tail.

This species is a rather more aquatic member of the family than the Common Newt, although it is still at ease on dry land, particularly outside the breeding season. It favours areas of still water with good weed cover as breeding sites; these can be anywhere from the bays of a reservoir to ditches and ponds. It tends not to stray too far from the breeding pools when it takes to land, hiding beneath logs and rocks.

Great Crested Newts feed at night, preying on water invertebrates and tadpoles. After a winter of hibernation (beginning in October), the newts emerge in March and take to the water. They spend the next five months in and around their chosen pool, where the female lays her eggs on aquatic plant leaves, just like the Common Newt. It takes around four months for the larvae to develop from egg to tadpole to adult newt. Sexually active from three years old, some Great Crested Newts have been recorded as more than 25 years old.

Common Frog

Rana temporaria Length: c.6–8cm; some females recorded at c.10cm

The Common Frog is a familiar sight to anyone who has rummaged around in the damper areas of their garden and has been taken by surprise by a frog leaping out of seemingly nowhere. Common Frogs are found across much of Europe (except for more northerly climes) and they are the commonest species on the continent. They can be seen in any moist, shaded area, be it a garden, a wood or field.

The Common Frog has relatively smooth skin and long hind legs – two features that immediately differentiate it from the Common Toad; although, in comparison to other European frog species, its legs are actually rather short. On closer inspection, its 'snouty' look is also different from that of the round-faced Common Toad, although Common Frogs do vary in this respect – as they age, their snouts become more rounded. Note too the large black eyes, surrounded by a golden orange eye-ring and dark flecking.

Grouped by frog aficionados into a collective known as 'Brown Frogs', the Common Frog is actually rather more variable in colour and tone than the collective name would suggest. The upper side can be anything from olive, brown or grey to straw yellow or even pink toned. Consistent features include the dark mask effect just behind the eye, the inverted 'A' mark on the top of the back (always black), along with black spotting (which is occasionally extensive) and the camouflage-like dark and light patches on the legs. The flanks are paler and rather marbled in appearance. The underside of the frog is generally white or yellow, but can be orange, and can show spots or marbling.

When breeding, the male Common Frog 'bulks out' his forelimbs, and his throat may develop a blueish hue. A breeding female also appears very large, with tiny white warts on her flanks. After mating, a female spawns, producing up to 1,500 eggs. After around three months, the tadpoles develop into froglets and they stay near water until hibernation time, in late autumn, when they bury themselves into pond mud. They reach sexual maturity at around three years of age.

Common Frogs feed on a variety of prey – worms, snails, beetles and slugs – all of which are flicked into the mouth by the long tongue.

Common Toad

Bufo bufo Length: 8–15cm, females always larger than males

The Common Toad is a large, rather plump, variably coloured toad which can live for up to 40 years. Found throughout Britain, Common Toads are absent from large areas of the Mediterranean, northern Scandinavia and Ireland. It hibernates from mid-autumn to late winter and on damp, mild February nights, dozens (even hundreds) of toads become active and move towards spawning areas.

The Common Toad has very warty skin that varies in colour from the typical dark brown to sandy yellow, olive, grey or even deep red. It is fairly uniform in colour, though it sometimes has darker blotches on the head and body. The top of the head, back and upper side of the flanks are frequently marked with dark green. The golden, or copper-coloured, eyes are not as prominent as those of the Midwife Toad and have horizontal, rather than vertical, dark pupils. Male Common Toads call at night, a gruff, repeated 'rarwk... rarwk... rarwk...' that doesn't carry over great distances.

Common Toads are found in a wide variety of habitats, but they usually favour damp areas of gardens and field edges, banks, and woodland fringes. Mainly a nocturnal hunter, the Common Toad feeds on spiders, various larvae, worms and insects. During the day, it often favours one hideaway, perhaps under a rock or a log, before emerging towards dusk. Like the Midwife Toad, the Common Toad moves with a shuffling walk or small hops.

The female lays her eggs two to four times a year, between March and August. The male positions himself on the female's back, and fertilizes the eggs as they are laid. The lines of spawn can be several hundreds to several thousands of eggs long, and after just 10 days or so, they begin to hatch and tadpoles emerge. The tadpoles' metamorphosis to toadlets takes around two to three months, and they leave their pond home in late summer.

Midwife Toad

Alytes obstetricans Length: no more than c.5cm

The Midwife Toad is a small, plump amphibian with prominent eyes. Its range was formerly from the Netherlands south-west to Spain, but it has been introduced to Britain, where it appears to be thriving.

The Midwife Toad has wart-covered skin, which varies in colour from the typical dark olive or dark brown to grey. The top of the head, back and upper side of the flanks are frequently marked with dark green. The almost protruding eyes are prominent and cat-like, complete with a vertical dark pupil. Male Midwife Toads call by night, a far carrying, resonant 'pooo... pooo... pooo...'. Birdwatchers who have been to the Mediterranean may recognise a resemblance to the call of the Scops Owl.

Midwife Toads are seen in damp areas of gardens, along field edges and in woodland. They are nocturnal hunters, choosing worms, spiders and insects as their prey. During the day, they often hide under logs, rocks and self-created small burrows. Like other toad species, Midwife Toads move with a shuffling walk or small springs across the ground.

The female lays her eggs from late winter (usually March) to late summer (mid to late August). After laying, the male wraps a string of spawn (which consists of between 20–100 fertilized eggs) around his legs and onto his rear end. He then carries them for around six to eight weeks, until they hatch.

Grass Snake

Natrix natrix Length: c.1–1.2m; some females recorded at over 1.8m

The Grass Snake is a large, shy, pale, thick-bodied snake with distinctive body colour and head markings. Found widely across England and Wales, the Grass Snake's range extends across Europe, to Russia in the east and to Spain in the south, as well as north into Scandinavia. It is absent from Scotland and Ireland.

Adult Grass Snakes have a noticeably rounded body, a rounded head and neat rounded eyes. The head and body are usually olive green in colour, with darker flecks along the length of the body. Some Grass Snakes can be greener, browner or greyer than the norm. Most distinctive of all is the yellow neck collar, edged with black. This, combined with the body colour and patterning, is a sure indicator that you are seeing a Grass Snake and not the venomous Adder.

Hibernating between late autumn and early spring, Grass Snakes mate during April and May. The females lay anywhere from 10-40 eggs in warm moist places, from compost heaps to haystacks. The eggs hatch during the late summer, and the young emerge as miniature replicas of adults. The male Grass Snake is sexually mature at three years of age, the female at four. Grass Snakes can live for up to 25 years.

Despite spending plenty of time on dry land, the Grass Snake is a very adept swimmer. This is reflected in its varied diet - tadpoles, frogs, fish and newts as well as mice, voles and birds.

Slow-worm

Anguis fragilis Length: c.30cm to just under 60cm

Despite its obvious resemblance to a snake, the Slow-worm is actually a legless lizard, with an un-snake-like fragile tail and eyelids that blink. A very smooth-skinned, slender reptile, the Slow-worm is also very shy and seen very infrequently. Slow-worms are found from Britain to Spain in the south, Russia in the east and Scandinavia in the north.

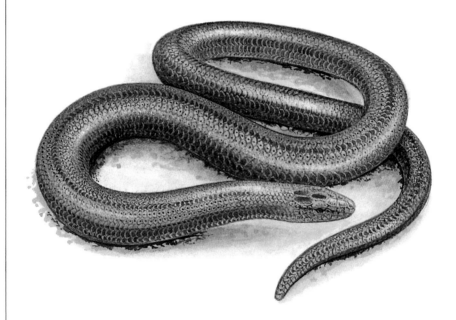

The tail of an adult Slow-worm is actually longer than the body itself, but often becomes damaged or broken and seldom grows back to its full length. Slow-worms are usually dark brown in colour, though they can sometimes appear reddish or coppery. The sexes are very similar, but the female can have a dark stripe down the middle of the back, and a dark belly and flanks. The male is more uniform. In contrast, young Slow-worms are more striking – golden brown above with dark sides, belly and back stripe.

Mating takes place between April and June. The young are born live after developing inside the female for three to five months. The female gives birth to as many as 12 young.

Slow-worms like plenty of ground cover and standing vegetation, and also damp conditions. They can be found in meadows, woodland clearings, hedgerows and heathland as well as gardens. Slow-worms are indeed rather slow, and will hunt fellow slow-moving invertebrates, often just after dawn or just before dusk. Because of their very secretive nature, it may be 'Lady Luck' that leads you to a Slow-worm, but they can occasionally be found beneath warm rocks or on old corrugated iron sheets.

Viviparous Lizard

Lacerta vivipara Body length: c.6cm; tail length: up to twice the body length

A scaly, long-bodied and long-tailed reptile, with a small rounded head and distinctive thick-necked appearance, the Viviparous Lizard is found right across Europe in well-vegetated and humid conditions. A garden with warm, dank areas of long grass is its ideal habitat. In Britain, it is also seen in woods and on moorland, heathland, and even dune slacks on the coast.

Viviparous Lizards have noticeably scaled skin, with a distinctive collar of raised skin behind the neck. Males tend to be larger than females, with a larger head and legs. The general patterning and colouration of both sexes is extremely variable. The commonest colour tends to be brown, but some can be grey or dull olive. The head, body and tail show varying degrees of flecking, and females tend to have a distinctive stripe down the back. The underside of the throat is white, as is the belly, although on some males it is orange, red or yellow. The underside also has many small black spots on most males and some females. Young Viviparous Lizards are always very dark.

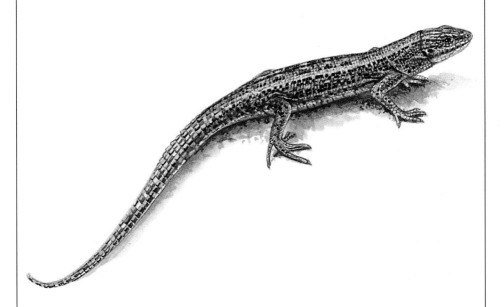

Favourite food items include insects, spiders, worms and snails, which it hunts during the day, using its excellent senses of sight and smell to track them down.

The female gives birth to live young between June and September. The young lizards develop within membrane-like 'eggs' inside the female. In some areas along the more southerly edge of its range, the female actually lays these eggs, and the young lizards hatch out soon afterwards. The Viviparous Lizard hibernates between October and March.

Chapter Seven
Birds

Birds are the most familiar of all animal classes seen in the garden. They belong to the class *Aves* and there are some 11,000 species around the world. Of these, around 750 have been seen in Europe, with roughly 560 species recorded in Britain and Ireland.

Almost every garden can attract many different bird species. If some food and water are made available, even a small inner city-garden can attract Starlings, Blackbirds, Robins and Wrens. If the garden is situated close to an area of green space, then other species such as the Great Tit or the Blue Tit may also appear. Small rural gardens will certainly attract the species listed above and even more, if the adjacent habitat is right. Coal Tits, Long-tailed Tits and Dunnocks are all likely, and even a woodpecker may pass through.

The majority of garden bird species are nest builders and the nestlings that hatch are entirely reliant on their parents for warmth, protection and food. The young birds develop quickly and many species produce a second clutch of eggs not long after the first brood have flown the nest. Some species, such as the Blackbird and the Blue Tit may have three to four broods, while recent milder winters have seen a sharp increase in some species breeding almost year round.

Grey Heron

Ardea cinerea Length: 84–102cm; wingspan: 155–175cm

The Grey Heron is a common resident breeding species in Britain and is seen throughout the year. As well as being a frequent visitor to gardens with ponds, it is found around park lakes, rivers, marshes and estuaries.

Grey Herons are early breeders, frequently displaying in January. They nest in trees, usually a short flight's distance from a pond, river or lake. Youngsters may be seen in the nest as early as mid- to late February.

Standing tall, slender and elegant, Grey Herons are imposing birds with a heavy, dagger-like bill. Their plumage is grey, white and black. In flight, they seem particularly big, flying slowly on bowed wings with their neck hunched in and their long legs trailing way beyond the tail.

The head has two broad black stripes over the back of the crown, which, in the breeding season, extend into slim black head plumes. The upperparts are blue-grey, with a distinctive black 'shoulder' and pale silvery plumes over the upper part of the closed wing. Juveniles (**far right**) always appear slightly stockier than adults and are darker, usually grey, but often admixed with brown.

Grey Herons are communal nesters, favouring parkland or woodland trees, but they will also breed in reedbeds and occasionally on cliffs. Their call is a loud, hard 'fra-ank' or 'kkrrank', heard in flight. Youngsters in the nest squawk and bicker constantly.

105

Sparrowhawk

Accipiter nisus Length: male 28–34cm, female 35–41cm; wingspan: male 58–65cm, female 67–80cm

The Sparrowhawk is a common, widespread breeding resident species and is seen throughout the year. In autumn, our residents are joined by migrants from the near continent and Scandinavia. The bane of garden bird life, the Sparrowhawk is a voracious predator. It moves with speed and guile, and is just as at home in city centre gardens catching House Sparrows as scorching through country hedgerows for Yellowhammers.

Male and female Sparrowhawks differ markedly in size and plumage, but both sexes share broad, blunted short wings, a small head and a long square-ended tail. The size difference is apparent when displaying birds are seen together in the early spring, the large, rather bulky female clearly outsizing the male.

In flight, look for the distinctive wing shape – the rather rounded wings almost form an 'S' shape. Sparrowhawks fly in quick bursts of rapid wing beats interspersed with short glides. When soaring, they look flat-winged and the tail is only occasionally fanned.

Males (**centre**) have blue-grey upperparts with dark wing-tips; their underparts are finely barred orangey-red from throat to belly. Females (**right**) have dark grey-brown upperparts except for a white supercilium and blackish looking bars on the tail. The underparts are white, with fine grey horizontal barring from the upper breast to the belly.

Common Kestrel

Falco tinnunculus Length: 31–37cm; wingspan: 68–78cm

The Common Kestrel is a widespread breeding resident in Britain and can be seen throughout the year almost anywhere, from cities to upland moors, and frequently along our endless road network. Although visits to gardens are scarce, these rather nervous birds of prey will snatch a meal if they spot one.

Common Kestrels are easily told from other birds of prey by their size, longish tail and noticeably pointed wings. Females (**bottom right**) are larger than males (**bottom left**) and are streaked warm brown above, buff below. The rufous mantle has dark bars extending onto the wing. The barred rump is slightly greyer brown, and the tail shows six or seven prominent dark bars. The male's head is blue-grey, with a dark moustache and buff throat. The mantle and wings are chestnut with black spotting and barring, while the rump and tail are pale blue-grey, except for the broad black tail bar.

In flight, the differences between the sexes are easy to see. The male's head and tail contrasts strongly with the wing pattern (**right**), while the overall rufous tones of the female (**left**) on the upperparts are clear.

The classic view of the Kestrel is with its head down and tail forward, flapping when necessary, hovering into the wind.

The commonest call heard from Kestrels is a shrill, piercing 'ke-ke-ke-ke'.

Pheasant

Phasianus colchicus Length: male 70–90cm (35–45cm of tail), female 55–70cm (20–25cm of tail)

The Pheasant is a widespread breeding species, resident throughout the year. Though generally found in woodland and on farmland, they do venture into country gardens. Introduced into some areas of Europe from Asia as many as 2,000 years ago, the Pheasant has been present in northern Europe for over 200 years.

Females (**near left**) are highly variable birds. Generally buff grey-brown all over with barring on the head, upper- and underparts, they are always smaller, slighter and shorter-tailed than males (**far left**). The spectacularly plumaged males also vary. The deep bottle-green head and 'ear tufts' contrast with the obvious patch of red facial skin. The mantle is a rich orange-red colour, blotched and spotted with cream, black and white. The long, spiky tail is warm brown, with coppery outer feathers and long, barred central tail feathers.

When disturbed, a male Pheasant flies off uttering an explosive, rapid fire 'kutuk, kutuk, kutuk'. Females have a higher pitched thin, reedy 'seeee-seeee'.

Moorhen

Gallinula chloropus Length: 27–31cm

The Moorhen is a common, resident, breeding species, seen throughout the year. Though most frequently seen around lakes, rivers, ponds and marshes, it is a frequent visitor to large, damp gardens, such as those that back onto a stream or nearby river. Although very happy on water, Moorhens actually spend a lot of time on land.

The Moorhen has a rounded head and body and short wings. The head and underparts are a dark greyish-blue, with browner upperparts. Along the flanks is a clear white line, which appears very bold against the dark wing and body feathers. The rear end is blackish, except for a gleaming white undertail. The frontal shield and bill (except for the yellow tip) are cherry red, as are the eyes. On land, the yellowy-green legs and large feet are obvious.

Young Moorhens have none of the adults' colour. The head is dark brown with a greyish wash on the face. The chin and throat are white, while the remainder of the underparts are washed grey-brown. The upperparts are brown.

Moorhens, although shy of humans, are quarrelsome birds and utter a loud, cursing 'kee-rrl' or sharp 'kik-keek' if their neighbours come too close.

Woodcock

Scolopax rusticola Length: 33–38cm (6–8cm of bill); wingspan: 55–65cm

The Woodcock is a relatively common bird, but rather hard to see. Although present throughout the year, across the whole of Britain and Ireland, autumn and winter are the most likely times for one to venture into a garden. It is a stocky bird, with a long bill and beautifully intricate camouflage plumage. It is actually a wader, and breeds in damp woodland and parkland (mixed or deciduous). The best time to see a Woodcock is at dawn or dusk on a spring day, when it may appear over a clearing in a wonderful, slow, slow, quick, quick, slow display flight.

In flight, the long bill, rounded wings and bright rufous brown upperwing are obvious. If disturbed, Woodcocks fly off quickly, low and straight; in less hurried circumstances, they fly more slowly. Look out for the barred underwing and belly. On the ground, the full intricacy of the Woodcock's markings can be appreciated. The brown, black and rufous tones merge to form a perfect camouflage. Look at the heavy barring on the crown and the contrasting rufous upperparts and barred buff underparts.

During the spring, 'roding' males utter a grunt-like, 'wark-wark-wark-yiup'. Autumn and winter birds flushed from a garden will invariably be silent.

Black-headed Gull

Larus ridibundus Length: 35–39cm; wingspan: 86–99cm

The Black-headed Gull is a common breeding species, seen throughout the year. Despite being a 'seagull', it breeds anywhere from coastal marshes to inland gravel pits, and is commonly seen in large numbers in autumn and winter, covering many miles as it flies off to roost. Braver birds steal scraps and crumbs from gardens, arguing raucously with each other as they do so.

Black-headed Gulls are slim-looking birds with a domed head, slender bill and clearly pointed wings. Both adults and juveniles have distinct summer and winter plumage.

A bird in its first winter (**right, top**) has mottling on the head and brown flecking on the wings. The bill and legs are a muddy orange.

An adult's winter plumage (**right, bottom**) is identical to the summer's, except for dark smudges behind the eye and duller bare parts.

First-summer birds (**below, right**) – approaching their first 'birthday' and therefore older than first winter birds – show a dusky hood and retain the brown wing and tail markings. The bare parts remain quite dull.

Adults in summer (**far left**) are conspicuous with their chocolate brown hoods. The upperparts are silvery-grey, with black flight feathers. The breast sometimes has a pink flush.

In flight, note the upperwing pattern of a first year bird (**top**). On adults (**bottom**), note the strong white leading edge, contrasting with the black trailing edge of the outer wing feathers.

The most frequent call is a raucous, rather harsh, drawn-out 'kreeare' or a hard, sharp 'kek'.

Feral Pigeon

Columba livia Length: 29–35cm; wingspan: 60–68cm

In many urban areas, and an increasing number of rural areas too, the Feral Pigeon is a very common breeding species, seen throughout the year. This familiar bird appears in an incredible variety of plumages, with colours varying from white to black, brown to grey, with many different combinations and patterns in between.

The Feral Pigeon has managed to infiltrate Rock Dove populations throughout northern parts of Scotland and Ireland, with a drastic effect on the wild stock. Genuine Rock Doves can still be found, but the task is growing harder. In any case, they only occur in western Scotland and Ireland, generally around coastal fields and cliffs.

The white rump and black wing bars are obvious. This bird (**near right**) is fairly close to the true Rock Dove. The blue grey head merges into a metallic green sheen on the neck, and the upper breast shows a purple gloss, contrasting with the greens and greys.

The Feral Pigeon's call is a monotonous, deep-sounding 'oooo-oooo', repeated over and over.

Collared Dove

Streptopelia decaocto Length: 31–34cm; wingspan: 49–55cm

The Collared Dove is a common and very widespread breeding species, seen throughout the year and in almost any garden. Although the species originates from Asia, by the mid-1900s it had embarked on a remarkable 'invasion' of Europe. It was first discovered in Britain in the early 1950s, in a small, secluded garden on the north Norfolk coast and, by 1956, a small breeding population was becoming established in East Anglia. Since then, the spread of the species has barely stopped and the bird is now one of the most familiar of all garden birds across the whole of Europe.

Collared Doves have broad wings and long tails. Adults are brownish on the upperparts and buff-pink below, with a distinctive black and white collar on the neck. The closed tail looks greyish, with buff and white outer feathers. Some Collared Doves are very pale, looking rather 'washed out'. Juveniles (**rear**) lack the neck collar and are scalier on the upperparts.

In flight, note the black wing tips (contrasting with the grey and brown inner wings) and the tail pattern, showing white tips to all but the central feathers, with blackish grey at the base. The call is a soft 'coo-coo-kut'.

Stock Dove

Columba oenas Length: 28–32cm; wingspan: 60–66cm

The Stock Dove is a relatively common resident breeding species in Britain, though it is absent from more northern areas. Favouring woods, parks and open farmland, Stock Doves do visit larger gardens but they are timid, especially if Woodpigeons or Collared Doves are present.

Stock Doves are smaller than Woodpigeons, with a small, squarish head, a plump body and beautifully subtle plumage. Unlike their larger cousin, they lack a white neck patch and have two short dark bars across the inner wing. They are dark grey above and some of the main wing feathers (the coverts and primaries) are black. The tail has a broad black tip. They have a dark silvery-grey head with an emerald green neck patch and a salmon pink breast. The underparts can be quite variable in colour, but are generally grey, sometimes looking rather blue. The bill is pale yellow, with a whitish 'knob' and reddish base.

The most commonly heard call is a low-pitched, monotonous 'coo-oh, coo-oh'.

Woodpigeon

Columba palumbus Length: 38–43cm; wingspan: 68–77cm

Woodpigeons are common and widespread across the country, and are found in many habitats. They are quite at home nesting in gardens, and will make frequent shuffling visits to the base of a garden bird table in search of crumbs. They are plump, full-chested birds with a small head, broad wings and a longish tail. In the autumn, huge flocks can be seen in agricultural areas, the numbers swollen by immigrants from the continent.

Adults (**far left**) are pale blue-grey above, with a greenish gloss on the hindneck, a white neck patch and blue-grey nape and rump. The breast is purplish-pink, fading to pale grey-white on the belly. In flight, note the white neck collar, large white wing patches and dark tail.

A juvenile Woodpigeon (**near left**) lacks the adult's white neck patch. The breast is duller and more buff-pink.

Woodpigeons may tend a brood of young squabs into the early winter if the weather is kind.

The call is a throaty, rather monotonous 'hoo-hrooah'.

Turtle Dove

Streptopelia turtur Length: 25–27cm; wingspan: 49–55cm

The Turtle Dove is a summer migrant to Britain, arriving from mid- to late April and leaving by early October. Formerly a relatively common breeding species, it suffered a dramatic decline in the mid- 1980s and it was only in the early 2000s that the numbers started to rise once more.

Found predominantly in parkland, woodland, plantations and bushy hedgerows, the Turtle Dove is a wary visitor to gardens, sitting unobtrusively on the edge of cover. It is a shy, small-headed dove with a slim body, long, graduated tail and colourful, intricately variegated plumage.

The head is grey with a buff-pink face. The neck shows several black bars with distinctive white edges. The wing is rich chestnut with bold black feather centres, with the exception of a grey forewing and dark flight feathers. The breast is pinkish grey, fading to off-white or pale grey on the flanks and belly. The upperwing shows grey, chestnut, black and white clashing in a flurry of colour.

Juveniles have a buff head, no neck bars and more subdued brown ring markings. The call is a soft, deep, purring 'rrrooooooorrr, rrrooooooorrr'.

Tawny Owl

Strix aluco Length: 37–43cm; wingspan: 81–96cm

The Tawny Owl is a widespread, fairly common breeding species in Britain and can be seen and heard throughout the year. Favouring open countryside and woodlands or parkland, Tawny Owls are also found in larger, wooded gardens across the country. They are, however, absent from Ireland.

The Tawny Owl is a mainly nocturnal, medium-sized, plump owl with a longish body. Its highly complex, intricate plumage varies from reddish brown to grey-brown, although in Britain the red phase is much more common; its upperparts are entirely warm brown with many black wormlike markings, while the underparts have a brown wash to the breast and flanks. 'Grey' Tawny Owls are rarely seen in Britain. Their plumage patterns are identical, with the brown tones replaced by ashy grey ones.

Juveniles (**top right**) have a striking, inquisitive look, their large black eyes peering out from a mass of down and feathers.

The call is the famous 'too-wit, too-woo' – *the* Owl call – but is actually closer to a long, drawn-out 'oo-ooo-hooo'. Listen too for a loud, penetrating 'kee-wick'.

Ring-necked Parakeet

Psittacula krameri Length: male 39–43cm (22–26cm of tail), female/young male 27–36cm; wingspan: 42–48cm

The Ring-necked Parakeet has established itself as a thriving colonist in the south-east of England, the result of a number of escapes during the 1950s and 60s. These medium-sized birds have startling plumage, slim wings and a long tail. Although they can be seen throughout the year, the best time to see them is in winter, when, in the late afternoon, they form massive roosting groups – often in hundreds. One famous winter roost near London has now grown to more than 1,500 birds.

Ring-necked Parakeets breed in wooded areas and parks, and as their population expands, so their range increases across the country. They are frequent garden visitors to bird tables and feeders, announcing their presence in a flurry of lime green and raucous squawking.

These are unmistakable birds, but note that the distinctive collar is present only on males. The bill is red, the tail is green and the underwing is also green.

Common Swift

Apus apus Length: 17–18.5cm; wingspan: 40–44cm

A trans-Saharan migrant, the Common Swift is a widespread summer breeding species that arrives back in Britain from mid-April onwards and leaves during late August or September. Common Swifts can be seen almost anywhere, from quiet country villages to the noisiest city.

With scythe-shaped wings, short forked tail and dark brown plumage, the Common Swift is an unmistakable aerial feeder. It is a fabulously competent flier, but on land it is cumbersome and awkward, shuffling on tiny legs and feet – this bird is built to fly.

The sooty brown tones of the upperparts are easily seen when Common Swifts fly close by. The trailing edge of the wing appears paler in some lights. In fast flight, the tail is closed, and the wings held swept back. When soaring, the wings are held further forward and the tail is open, giving an altogether different silhouette.

Juveniles appear browner and scalier than adults, with a more extensive white throat bib.

The call is a piercing, shrieking, repeated 'skreeeeeeee-skreeeeeeee' – one of the unmistakable sounds of summer.

Green Woodpecker

Picus viridis Length: 30–36cm; wingspan: 45–51cm

A fairly common resident breeding species, the Green Woodpecker is seen in woodlands, open parkland and gardens throughout the year, though it is absent from northern Scotland and the whole of Ireland.

Long and robust-looking, Green Woodpeckers have a sharp, powerful bill, large grey feet (two toes forward, two toes back) and a short, spiky tail that is used in a crampon-like style on tree trunks. Despite their undoubted skill on trees, you may also see one hopping purposefully on a lawn, busily searching for ants and grubs.

Green Woodpeckers' flight is undulating, with deep beats followed by closed wings as the bird progresses. Approaching a tree, the Woodpecker arrives with a huge upward sweep before landing on a trunk.

The male (**far left**) has a red centre to the 'moustache'. Both sexes have a dull yellow rump and uppertail, while the graduated tail is brown-grey with darker bars. On the female (**near left**), the 'moustache' is wholly black. The cheeks and throat are pale greyish white to pale yellow. The call is an unmistakable, laughing 'yaah-yaah-yaaah' – the famous yaffle.

Great Spotted Woodpecker

Dendrocopos major Length: 23–26cm; wingspan: 38–44cm

The Great Spotted Woodpecker is a widespread, common resident species in Britain, with numbers topped up in the autumn, along northern and eastern coasts, by occasional migrants from Scandinavia. It can be seen in parks, woods and gardens around the country, but is absent from Ireland.

Like other woodpeckers, the Great Spotted Woodpecker has a thick pointed bill, graduated tail, 'reverse' toes and undulating flight.

Males (**far left**) have a red nape, which is absent on the female (**near left**). Both sexes share bold black and white patterning on the upperparts, along with bold red patches on the undertail. Note the longer central tail feathers, essential for balance when the bird is on a tree.

In flight, Great Spotted Woodpeckers are particularly striking. The bold white shoulder patches and white spotting on the wings really stand out. Juveniles are similar to the adults, with red crowns and more sullied underparts.

The call is an explosive, excited 'chik'. In spring, males have a very loud, rapid drum, faster than any other woodpecker.

115

Lesser Spotted Woodpecker

Dendrocopos minor Length: 14–16.5cm; wingspan: 24–29cm

The diminutive Lesser Spotted Woodpecker is a locally scarce resident breeding species. Present throughout the year, it is best looked for in late winter and very early spring. With no leaves on the trees, the male becomes ever more visible, and audible, as the search for a partner begins.

At around the same size as a House Sparrow, this is Europe's smallest woodpecker. It is generally seen in the deciduous woods and parkland of southern England, and is an increasingly familiar visitor to gardens. It is absent from Scotland and Ireland.

The Lesser Spotted Woodpecker has neat pied plumage with a small sharp bill, and the familiar graduated tail and reverse toes of the rest of the group. Juveniles (**left**) resemble adults (**below**), with a whitish forehead, flecked black, spottier flanks and a buff wash to the face.

Males (**below left**) have a neat red crown compared with the whitish crown of the female (**below right**). The upperparts are barred black and white, with off-white underparts. There is no red on the underparts, unlike the Great Spotted Woodpecker.

'Lesser Spots' have a typically undulating flight, during which the barring on the back and wings is very striking. They are able to feed on smaller branches and twigs than other woodpeckers – sometimes very tiny indeed.

The call is an excited, rather Kestrel-like 'kee-kee-kee'. The drumming is *weak* compared to that of the larger and more powerful Great Spotted Woodpecker.

Swallow

Hirundo rustica Length: 17–21cm (tail projection 3–6.5cm); juvenile 14–15cm

Along with the Cuckoo, the Swallow is the avian herald of spring. A common widespread breeding species, this summer migrant clocks up tens of thousands of miles a year and arrives here from early to mid-April onwards, generally leaving to head south by early to mid-October.

Swallows can be seen almost anywhere, from coastal marshes to urban streets, and their nests are a familiar sight under the eaves of houses. Male Swallows (**below right**) are identical to females except for their longer tail streamers. As the breeding season progresses, these long streamers are often lost.

These graceful fliers have a white forewing, pale belly and white spots on the underside of the deeply forked tail. They can be separated from Swifts and House Martins by their size, more pointed wings, manner of flight and obvious tail length.

Juvenile Swallows (**far left**) are duller than adults. Very young fledglings show flecks of down on the head and a strong yellow gape.

The call is a tinkling 'vit vit' or occasional 'splee-plink'. The male's song is a strong, fast, twittering warble.

House Martin

Delichon urbica Length: 13.5–15cm

As with the Swallow, the House Martin is a well-travelled summer visitor, clocking up thousands of miles every spring and autumn. It is a common, widespread breeding summer migrant, arriving from late April onwards and leaving by October, though a few linger late into the year.

House Martins can be found across the country, but they always nest under the eaves of a house whatever the location. They return, year after year, to the same building.

Clearly smaller and more compact than Swallows, House Martins have broad wings, a shortish notched tail and almost pied looking plumage.

Juveniles (**left**) are dull version of adults (**right**), appearing browner on the head and wings, and duller blue on the upperparts. The underparts are off-white with a dingy wash to the throat. Very young fledglings show a yellow gape at the base of the bill.

The call is a dry rolling 'preeet-preet', often repeated.

Pied Wagtail

Motacilla alba yarrellii Length: 16.5–19cm

The Pied Wagtail is always busy, always fearless and always fun, and is a common, widespread breeding species that can be seen throughout the year, although some are spring and autumn passage migrants.

Pied Wagtails have striking plumage, a thin bill, rounded body and long tail. They are very approachable, with little fear of humans; they can be seen in almost any garden, marching across the rooftops, jumping up for flies on the lawn or coming to window feeders for cheese, which they love.

Males (**left, top**) have entirely black upperparts, with a white forehead and white cheeks. The black wings show obvious wingbars. Note the large black bib, extending from the base of the bill to the upper breast. The female (**left, bottom**) has a greyer back and less black on the head and breast.

Juveniles (**below left**) have greyish-brown upperparts and head, the underparts are buff with an indistinct black bib.

On occasion, you may see the continental version, the White Wagtail (**below**). These are migrants, and have pale grey upperparts, contrasting strongly with the head pattern. Female White Wagtails are duller, with less black on the head.

The call is a high-pitched 'chiz-zik' or 'slee-vit'.

Waxwing

Bombycilla garrulus Length: 18–21cm

As subtle as they are beautiful, Waxwings are scarce winter visitors to Britain. In years when there has been a crop failure in their usual Scandinavian or Russian wintering grounds, these glorious birds migrate en masse to more westerly or southerly countries in search of berries. These winter invasions are known as 'irruptions'. In good irruption years, thousands of birds make it to Scotland and the north-east of England, then often drift further south to both inland and coastal areas.

Usually appearing here from mid- to late October onwards, Waxwings can continue to arrive throughout the winter. They leave us to head back north or northeast in late winter or very early spring.

This bird's favourite food is a variety of red berries. The planning of many housing estates and out-of-town shopping centres has seen numerous berry bushes planted to profile a site, and these are often hugely popular locations for Waxwings.

Waxwings have velvety pinkish plumage with a distinctive head crest and a jet black eye mask and bib. The waxy red tips on the secondary feathers are quite pronounced, and also lend the bird their name. Flying Waxwings are reminiscent of a slim Starling, with broad wings, and a shortish square tail. Note the pale underwings, rufous vent and grey upperwings.

Females (**far left**) are similar to males (**near left**) but are slightly smaller, with a shorter crest and a less well-defined sootier bib. The waxy red tips are less obvious, as are the yellow tail band and primary tips. Young Waxwings have shorter crests and browner plumage.

The call is a delightful, gently whistling 'ssirrrr', which can reach quite a crescendo in a large flock.

Wren

Troglodytes troglodytes Length: 9–10.5cm

A common, widespread resident breeding species seen year round, the Wren is a versatile little bird – as well as being a fan of garden bushes, it can be seen in parks, any area of scrub, farmland, reedbeds, and even (in the northern and western islands of Scotland) on cliff faces.

Wrens are tiny birds with a stubby cocked tail and rich brown plumage, finely streaked above with barred outer wings and buff underparts. They are secretive by nature and can be hard to see, furtively seeking food in tangled undergrowth, but you may spot one perched with tail cocked as the bird dips up and down, churring. Their nests are tiny rounded affairs, often tucked away in brambles or a hedge. A male Wren makes several nests for females to choose from.

In flight, often all you see is a tiny brown bird whizzing by on whirring wings.

The call is a rapid-fire 'cherr, cherr' cherr', or explosive melodic trilling.

Dunnock

Prunella modularis Length: 13–14.5cm

The Dunnock is a common resident breeding species, seen year round. A remarkably secretive and rather unobtrusive bird, it loves scrubby cover, hedgerows and gardens across the whole country.

Although superficially resembling a sparrow, the Dunnock is more slender than the House Sparrow and has a thinner bill, a longer tail and rather different plumage. Adults are warm brown above with distinctive grey faces. The upper- and underparts are brown, streaked with black. The sexes are similar, but males tend to be greyer on the head and throat.

Juveniles are duller and streakier than adults. The head is much browner and the underparts are heavily blotched.

The male Dunnock has a delightful display routine to attract a female. Landing on a small branch, he flicks one wing then the other, in turn, for several minutes. One lone male will quickly draw a crowd, and the usually solitary Dunnock will be joined by a group of like-minded wing-flickers.

The call is a thin 'teeh'. The song is resonant and, for such a dowdy bird, sweet and rather melodic.

Robin

Erithacus rubecula Length: 12.5–14cm

The Robin is perhaps *the* garden bird. A common, widespread breeding resident species, it is seen throughout the year and loves gardens that border woodland or scrubby areas. In autumn, numbers are swollen with the arrival of continental migrants.

With its plump shape, red breast and the lovely song of the male, the Robin is one of our most recognisable species. The bold orangey-red face and breast are bordered with grey underparts, while the upperparts are a contrasting brown.

Robins often begin to sing in the winter, proclaiming their territories from a very early stage. They are aggressive and will fight, chase and harry each other to maintain their patch.

Juveniles are just as distinctive as adults. The head, upperparts and breast are liberally speckled with dark brown spotting and scalloping, which sometimes extends onto the flanks. Young Robins acquire red breasts from June to September.

The call is a sharp, hard 'tik', the song a blend of excited, melodic tumbling notes.

Black Redstart

Phoenicurus ochruros Length: 13–14.5cm

The Black Redstart is primarily a city garden bird. It is a scarce breeding species in Britain, favouring disused buildings, chimneys and factory roofs. It is found mainly in the southeast (London in particularly famous for its 'Black Reds'), but it is also a passage migrant, primarily in the autumn, when it may be seen almost anywhere in the country. The Black Redstart is a rather slim member of the Chat family, with a habit of perching bolt upright and shivering its tail.

Breeding males (**near left**) are handsome, with a jet-black face and upper breast contrasting with slate-grey upperparts and belly. The wing shows a prominent white flash. A non-breeding male (**bottom left**) resembles a female, but is generally darker and greyer. The tails can look very dark, as is the case here.

Female and young Black Redstarts (**top left**) are sooty brown all over, except for darker wings and the orangey-red tail, which often looks rather dark. Note the fine black bill and button eye.

The call is a hard-sounding 'tuk' or a whistled 'wist'.

Blackbird

Turdus merula Length: 23.5–29cm

The Blackbird is a widespread breeding, resident and migrant species. They are very common, seen year round, and are big fans of gardens as well as parks and woodland across the country. In autumn and early winter, migrants arrive on north and east-facing coasts.

Blackbirds are typical thrushes – stocky, with plump bodies, rounded heads, longish wings and a longer tail. The male (**far left**) is the only jet-black bird of its size seen in the region. Note the bright yellow bill and eye-ring.

Adult females (**near left**) vary widely in colour from rufous to grey-brown. The throat is well marked, and note the dull bare parts, similar to a juvenile. Juveniles appear more rufous than adult females and they are more heavily mottled on both upper- and underparts. The throat and breast can look rather speckled and thrush-like.

Blackbirds have a wide range of calls – a nervous 'chink-chink' (often heard at dusk) is the most common sound, this or an urgent, repeated alarm call. The male has one of the loveliest songs of all birds – slow, cheery and fluting.

Song Thrush

Turdus philomelos Length: 20–22cm

A widespread species, the Song Thrush is a common resident across Britain and Ireland. From September onwards migrants arrive from the continent, mainly on the north and east coasts, often in some numbers.

Song Thrushes breed in dense garden hedges, woods and parkland. Clearly smaller and more compact than the Mistle Thrush, they have shorter tails and lack obvious pale fringes on the wing, and the breast is less boldly marked. The brown upperparts contrast with yellow washed spotty underparts; there is a hint of a supercilium on the head that the Mistle Thrush lacks.

In flight, the spotting on the underparts can be seen clearly. Check for the orange on the underwing, the only common thrush to show this pattern. The call in flight is a thin but far-carrying 'tssip'. The alarm call is a loud 'chuck-chuck'. The song is clear, languid and fluty, with many mimicked and repeated phrases. Once one phrase has been sung three or four times, a new phrase takes over.

Fieldfare

Turdus pilaris Length: 22–27cm

A fairly widespread winter visitor, the Fieldfare is one of the most elegant and richly toned birds you'll see in your garden. Arriving from northern European breeding areas, Fieldfares stream into the country from mid to late September, with the bulk of them appearing in late October and early November. They will usually leave by late March, but a few often linger into April. A few pairs, very occasionally, remain in northern uplands of England and Scotland to breed. They can be seen in gardens almost anywhere during the winter, feeding on berry bushes or windfall apples – a particular favourite.

Large and rather bulky, the Fieldfare is reminiscent of the Mistle Thrush in size and shape, although the contrasting maroon, grey, yellow, white and black plumage and distinctive call make the Fieldfare an easy bird to identify. Face on, the head and breast patterns contrast markedly; from the side, the maroon back and wings contrast with the grey rump and back.

The sexes are similar, though the female Fieldfare (**below left**) has finer black streaks on her crown and is generally duller than the male (**below right**).

In the winter, Fieldfares commonly associate with other thrushes (particularly Redwings), forming large roving flocks that can quickly strip a berry-laden hedge.

The call is a loud, chattering 'chak, chak, chak'.

123

Redwing

Turdus illiacus Length: 19–23cm

Redwings are much like Fieldfares in movements and general habits. A widespread winter visitor from northern Europe, they arrive from the middle of September onwards and head back to Scandinavia, Iceland and beyond from the late days of winter onwards. Very occasionally, they set up territory in northern hills here to breed. Again, like the Fieldfare, the Redwing is very partial to berries and windfall apples and often gathers in disorderly flocks.

Redwings resemble Song Thrushes but are smaller and darker, with a whitish supercilium and 'moustache' contrasting with a brown head and upperparts. The underparts are white, with dark slender streaks extending to the flanks. Most obvious is the vibrant red flank patch, which extends under the wing.

In flight, the clearest feature is the red underwing and flanks. Close up, the head pattern and streaks below may also be seen.

The call is a thin, high-pitched 'tzzip', heard when disturbed from cover or, on clear winter nights, as a contact call between newly arriving birds.

Mistle Thrush

Turdus viscivorus Length: 22–27cm

The Mistle Thrush is a relatively common resident breeding species in Britain, as well as being a scarce migrant. Seen throughout the year, it likes gardens that either border coniferous woodland or have evergreen trees within them. It also likes large open spaces and, occasionally, large groups of a dozen or more will chase excitedly around. For such a strident, confident bird, the Mistle Thrush can be rather timid in a garden.

Mistle Thrushes are large, boldly marked birds with a long body and tail, and are relatively easy to distinguish from Song Thrushes. Mistle Thrushes can look rather 'pigeon chested', particularly when the wings and tail are held drooped, nearly at ground level. The bold spotting and cold grey tones of the upperparts are obvious, but also note the double wing bar too. The spotted breast shows a yellow wash on the flanks. In flight, note the white underwing, as well.

The call is a loud, wooden 'zrtt-r-r-r-r-r', like a football rattle, often made in flight. A male Mistle Thrush will almost always head to the top of a tree to sing, and begins singing in late February or early March. The song is similar to the Blackbird's, but tends to be more raucous, quicker and a touch more monotonous.

Blackcap

Sylvia atricapilla Length: 13.5–15cm

The Blackcap is a secretive but common breeding migrant species. Over the past two decades, with the slight rise in global temperatures and generally mild winters in Britain, many have taken to wintering here, but most arrive in mid-April and leave by mid-November.

Favoured summer habitats include scrubby woodland, dense hedgerows and thick garden cover across most of the country. As winter approaches and insects become scarcer, Blackcaps often move into gardens and can be seen on bird tables, scrapping with the Blue Tits and Greenfinches.

The Blackcap is larger and stockier than the Chiffchaff or Willow Warbler. Males (**bottom**) are unmistakable with their glossy black cap that contrasts with silvery cheeks and nape. The upperparts are entirely silvery-grey, darker on the wings and towards the tail tip. The underparts have a greyish wash on the breast and flanks. Females (**top**) have a rich russet-brown cap, a browner wash to the underparts, with browner wings and a greyish-brown wash to the underparts.

The call is a hard 'teck' or 'tac', typical of birds in the *Sylvia* group. Blackcaps are not renowned for singing from treetops, preferring to advertise themselves from the cover of a hedge or other tallish vegetation. The song is loud, very melodic and far-carrying.

125

Chiffchaff

Phylloscopus collybita Length: 10–12cm

Like the Blackcap, the Chiffchaff has adapted to life in our climate – once a common and widespread summer migrant and breeding species, the Chiffchaff can now be seen across the country throughout the year. In autumn, birds arrive from Scandinavia and Siberia, bolstering the transient British population; they are generally greyer brown than the more olive-brown residents, but it takes an experienced eye to pick them out.

Chiffchaffs like any type of woodland habitat in springtime, and often take up territory in trees or scrub in larger gardens. In winter, they appear more frequently in gardens, searching for insects in woodpiles or flying midges on milder days. They are also fond of searching for flies and midges at sewage treatment works, so if you live near one, pay it a winter visit!

The Chiffchaff is slightly smaller than the very similar Willow Warbler, but has a more rounded head, shorter wings and a stockier appearance. Note the greyish-white supercilium, blackish eye-stripe and pale looking ear coverts. The upper-parts are generally olive-green, with darker wings. The underparts are buff-white except for a white throat. The thin bill is darkish. The legs and feet are dark brown, unlike the Willow Warbler's.

In spring, Chiffchaffs move quickly through cover, constantly flicking wings and jerking from branch to branch.

The call is a plaintive 'hueet'. The song is a repetitive 'chiff-if-chaff', heard from late winter onwards.

Willow Warbler

Phylloscopus trochilus Length: 11–12.5cm

The Willow Warbler is a common and widespread breeding summer migrant, arriving in early spring and leaving throughout the autumn until late October. Favouring areas with young trees and bushes or dense garden scrub, though Willow Warblers rarely pay more than a flying visit to most gardens.

Willow Warblers bear a strong resemblance to the stockier Chiffchaff, but have slightly longer wings, paler legs and different overall plumage tones. In spring, they appear generally paler than Chiffchaffs. The head and upperparts have a long, yellow-tinged supercilium, not short and buff like that of the Chiffchaff. A pencil-thin black eyestripe and slightly blotched cheeks are also apparent. The underparts are cleaner than the Chiffchaff's, lacking buff tones. The bill shows a dark upper mandible and tip, with a flesh-pink lower mandible. The legs are pale.

A juvenile in autumn can be particularly striking. The upperparts are pale olive-green, with rich lemon underparts.

The call is similar to the Chiffchaff's, but is a more penetrating 'hoo-eat'. The song is a lovely, simple descending set of flourishing notes.

Goldcrest

Regulus regulus Length: 8.5–9.5cm

The Goldcrest is Britain's, and indeed Europe's, smallest bird. It is a widespread resident breeding species, and also an autumn migrant to our coasts. Migrants can arrive in droves – tiny, tired bundles that are so exhausted they may land on you, hoping to find shelter on a human 'tree'.

Goldcrests can be found anywhere that boasts some coniferous tree cover – gardens, parks and woods are all favourites. They are agile birds, flicking constantly for aphids or flies. When not moving through treetops, they are equally at home foraging through small bushes and grass.

Males (**right**) have rich red central crown feathers, contrasting with black borders, and greenish upperparts. They have black and white wings and a short, dark, notched tail. The female (**left**) resembles the male, except for a bright yellow crown stripe.

The call is a high-pitched, far-carrying 'zee-zee-zee-zee-zee'. The song is a continually repeated, squeaky 'peet-ee-liu peet-ee-liu', reminiscent of an unoiled bicycle.

127

Spotted Flycatcher

Muscicapa striata Length: 13.5–15cm

The Spotted Flycatcher is a relatively common and widespread summer migrant, absent only from north-western Ireland. One of the later spring migrants, it appears in early to mid-May, and tends to leave our shores by the middle of September. From the early 1990s, numbers declined dramatically, but they are now rising again.

Favouring parkland, open woodland and well-vegetated (though not necessarily large) gardens, the Spotted Flycatcher is a sparrow-sized bird with long wings, a squarish tail, short legs, big eyes and a broad-based bill. Adults have grey-brown upperparts, with some dark streaking on the head. The wings are darker, but note the indistinct white bars. The underparts show streaking on the throat and breast. The dark tail has white outer feathers.

Juveniles (**right**) have a heavily scalloped, scaly-looking head, mantle and breast, contrasting with the broadly fringed grey-brown wings and dark tail.

In flight, the Spotted Flycatcher is very acrobatic. It twists and turns close to the ground before returning to the same twig or branch, catching assorted flying insects including small bees, butterflies and greenfly.

The call is a thin 'tzee'. The song comprises several wheezy notes, repeated frequently.

Long-tailed Tit

Aegithalos caudatus Length: 13–15cm (including tail, 7–9cm)

The Long-tailed Tit is a common resident breeding species, and also an occasional continental migrant. It is absent from northernmost Scotland and Ireland. It is found year round in gardens and any woodland habitat, and is most obvious in winter, when it forms large, vocal flocks.

The tiny oval body, rounded head, short stubby bill, distinctive plumage and massively long tail make this species unmistakable. They have whitish heads with two broad black bands running from the bill base to the nape. The wings show pink scapulars (the feathers on the birds 'shoulders') and broad white edges to the tertials (the long, overlying feathers near the wingtip). The underparts are off-white, with a pink flush to the flanks and belly. The long tail is black with white outer feathers, spotted white on black from below.

Unlike other tits, 'Long-tails' prefer eating insects to seeds. Highly agile in their search for food amongst leaves and twigs, they hang upside down by one foot while clutching food in the other.

The delicate oval nest (**above**) created by the adult birds is a fragile work of art and is quite unlike that of any other British bird.

The call is a far-carrying, piercing, squeaky 'tsee, tsee, tsee'.

129

Marsh Tit

Parus palustris Length: 11.5–13cm

The Marsh Tit is a relatively common, resident breeding species seen throughout the year. It favours damp areas of broad-leaf woodland, but will come to garden feeders if its favoured habitat is close by. It is absent from Ireland and most of Scotland.

Marsh Tits have thick necks, a stubby bill and a small, rounded head. They are very similar to Willow Tits, but the wings are plain, lacking a wing panel. In early spring, they may show some pale edges, but they will always be less obvious than a spring Willow Tit. They have glossy looking caps, extending to the rear nape. The cheeks are white, fading to buff behind the ear coverts. Notice the small black bib on the chin.

Marsh Tits are frequent users of nest holes. Unlike Willow Tits, they do not excavate their own, preferring to make use of natural cavities.

The call is a nasal 'pitchou' or 'pitchou ke-ke-ke'.

Willow Tit

Parus montanus Length: 12–13cm

A scarce but resident breeding species, Willow Tits are seen throughout the year but are less common garden visitors than Marsh Tits. They are absent from Ireland and most of Scotland and share their cousins' liking for damp woodland, particularly of alder or birch.

Willow Tits have a particularly thickset appearance, with a large head and a noticeable 'bull neck'. They have a dull black cap extending to the mantle. The cheeks are wholly white (unlike the Marsh Tit's white and buff) and the bib on the chin is larger. Fresh plumaged Willow Tits have a distinctive pale wing panel, lacking on the Marsh Tit.

Head on, the large-headed look is more apparent, as is the dull, sooty crown. The black bib (which resembles a moustache here) is always larger and broader, and the slightly rounded tail has paler outer tail feathers than the Marsh Tit's. Note too, the warmer tone on the flanks.

Willow Tits nest in holes which they excavate themselves in rotten tree stumps – unlike Marsh Tits, which, as already mentioned, use natural holes.

The call is a loud, resonant, buzzy 'tchay-tchay-tchay'.

Crested Tit

Parus cristatus Length: 10.5–12cm

The Crested Tit is an enigmatic, locally scarce breeding resident, confined to the Caledonian forests and gardens of the Scottish Highlands. Frequently seen feeding in native pines, 'Cresties' adopt many of the familiar family postures and traits.

A rather small member of the Tit family, they can, on occasion, be rather tricky to see. Equally though, they can be really inquisitive and approach to within touching distance.

Unmistakable birds, Crested Tits have grey brown upperparts, buffy toned underparts and a striking black and white head pattern, topped off by a chequered black and white crest. Fabulous!

Its most frequent call is a buzzing 'buurrrret'.

Coal Tit

Parus ater Length: 10.5–11cm

A common breeding resident, and migrant, species, the Coal Tit is seen year round. In the autumn, birds from continental Europe arrive, sometimes in considerable numbers. Coal Tits are frequent visitors to gardens, especially those with coniferous borders.

This is a curiously proportioned bird, with a large head, short forked tail and thin bill. Adults (**below**) have glossy black crowns and bibs, contrasting with bold white cheeks and a rear nape patch. The upperparts are olive grey, with prominent white wing bars. Below the bib is a small white patch on the upper breast that fades to pinky-buff.

Irish Coal Tits have a yellowy wash to the nape and cheeks, and the upperparts are buff in tone with yellow-buff underparts. Continental migrants from northern Europe differ from British birds in having whiter cheeks, a greyer back and paler buff-pink underparts.

Juveniles (**above**) are rather drab. The basic patterns are the same as the adult, but the colours are more subdued and appear yellow all over.

The call is a thin 'tseu', the song a distinctive 'pitchou, pitchou, pitchou'.

Blue Tit

Parus caeruleus Length: 10.5–12cm

One of the commonest and most instantly recognisable garden birds, the Blue Tit is a widespread breeding species, seen throughout the year. It is also seen in woods, parks and almost any hedgerow across the country. In autumn and winter, the flocking instinct takes over and family groups join together to rove around with other members of the Tit family (especially Long-tailed Tits), Treecreepers and Goldcrests.

Adults (**left**) have sky blue crowns, contrasting strongly with the white and black of the rest of the head. The mantle and rump are lime green, and the slightly notched tail is blue-grey. The wings are sky-blue with a thin white wing bar. The underparts are lemon yellow, with a black central streak on the belly.

Youngsters are basically yellow, green and black and appear duller than adults, with no trace of white. They often sit close together soon after fledging.

The call is a rather cheeky, harsh 'churr-urr-urr'.

Great Tit

Parus major Length: 13.5–15cm

The Great Tit is a common resident breeding species favouring gardens, parks and woodland throughout the year. The largest member of the family seen in Britain and Ireland, Great Tits are almost House Sparrow size, and can be rather aggressive.

Males (**left, bottom**) have a glossy black head with bold white cheeks and a yellowish nape. The upperparts are brighter olive than the female's. The belly is bright yellow, with a broad black central band. Females (**left**) are noticeably duller than males. The head looks duller and the mantle tends to be paler green. The underparts are yellow, but with a narrower central black stripe.

Juveniles (**right**) look washed out compared to adults. The plumage is subdued, with yellow on the cheeks and underparts. The black breast stripe is indistinct.

The call is a loud, resonant 'teecha-teecha-teecha' and a ringing 'zinc, zinc'.

Nuthatch

Sitta europaea Length: 12–14.5cm

A relatively common resident breeding species, the Nuthatch is a frequent visitor to gardens throughout the year. Its favourite habitat is deciduous, mature parkland, but it is also happy raiding a peanut feeder. The Nuthatch is found south of a line running from the Solway to the Tees, but is absent from Ireland.

The Nuthatch has a compact, torpedo-shaped body, short tail and hefty, sharp bill; its upperparts and wings are steely blue-grey, its tail sides black and white. Note the broad black eye-stripe, contrasting with the white throat. Males have chestnut flanks, which are paler on the female. The bill is blackish, with a silvery base. Young birds have brown-tinged upperparts, a narrower, duller mask and dull brown flanks.

Thanks to its size, shape and undulating flight, the Nuthatch resembles a small woodpecker, but a closer look reveals the grey upperparts and buff-orange underparts.

The call is a loud, full-sounding 'chewit-chewit'.

Common Treecreeper

Certhia familiaris Length: 12.5–14cm

Seen throughout the year, the Common Treecreeper is a common breeding resident and migrant species. Rather secretive in the spring and summer, when they nest in woods and parks, Common Treecreepers become a little more obvious during the winter as they join roving tit flocks in search of food.

This is a small, almost mouse-like bird, with a short, curved bill, beautifully camouflaged plumage and elusive habits. It has mainly dark brown upperparts, boldly blotched with white. A long, white supercilium contrasts with darker cheeks. The tail has a very distinctive shape, due to the longer, pointed central feathers, which enable the bird to 'clamp' onto a trunk. The wings are beautifully and intricately marked, with black, creams, buff-yellow and grey-browns. The underparts are silky white, with a faint buff wash to the flanks.

When seen in its undulating flight, note the Common Treecreeper's buff and black wing bars, rufous rump and notched tail.

The adult finds food by moving up and around the trunk and branches of one tree, picking up insects from the bark crevices, then dropping to the bottom of another and working its way up.

The call can be tricky to hear, but is a thin, high-pitched 'tsee'.

133

Jay

Garrulus glandarius Length: 33–36cm; wingspan: 54–58cm

The Jay is a shy but colourful resident breeding species, seen year round. In autumn it may be joined by migrants from Scandinavia and Russia, occasionally in very large numbers. Jays are absent from northern Scotland and south-west Ireland.

Favouring woodland fringes, plantations and parkland, the Jay is a frequent visitor to larger, open gardens. It is fairly chunky with a stout, heavy bill; mainly pinkish-buff with a streaked crown, the Jay has a distinctive face pattern and striking black and white wings with a beautiful aquamarine 'elbow'. The rump is snowy white and the longish tail is black.

On the wing, Jays are dazzling – pink, black, white and blue merge in a flight that is slow, deliberate and undulating. In autumn, Jays congregate in small groups to collect and store nuts and acorns for winter food. They hop clumsily over the ground, and may bury their food if they find enough.

The call is a loud, harsh 'kraa'.

Magpie

Pica pica Length: 40–51cm (including tail 16–20cm)

Equally at home in a large garden or coastal bushes, woodland or moorland, the Magpie is a common resident breeding species across the whole of Britain and Ireland, with the exception of northern Scotland. Its black, white and iridescent plumage, chunky, stout bill and long black tail make this bird unmistakable.

Magpies have entirely jet-black upperparts and breasts. The belly is bright white, contrasting with the black undertail. Note the bold white shoulder patch. The long tail has a gleaming green iridescence, turning to blue at the tip.

The Magpie flies with a combination of quick beats interspersed by glides. The white flight feathers and the prominent white 'V' on the back contrast strongly with the rest of the dark plumage. Note too the long, tapered tail.

The call is a loud, chattering 'chacka', and also a hoarse 'tzsee-tzsee-tzsee'.

Jackdaw

Corvus monedula Length: 30–34cm; wingspan: 64–73cm

A relatively common resident breeding species, and occasional continental migrant, the Jackdaw can be seen throughout the year. An adaptable bird, the Jackdaw is as at home in the cities as it is in the countryside, though it is furtive in gardens. The species is often seen perched on chimney stacks.

The Jackdaw is a short-billed, compact member of the Crow family. The head has a glossy black cap and bib, contrasting strongly with a pale grey nape. The upperparts are sooty black with a strong green-purple gloss in certain lights, particularly on the wings. The underparts are dark grey. Note the white eye and short, thick, silvery black bill.

When flying to roost, Jackdaws put on a remarkably agile aerial display; birds can drop like a stone out of the flock, and more will follow.

The call is a strident 'keyaa' and 'ky-aack'.

Rook

Corvus frugilegus Length: 41–49cm; wingspan: 81–94cm

The Rook is a common resident breeding species found throughout the year in Britain and Ireland, though it is absent from northern Scotland. Rooks breed in open parkland, farmland and woodland. They are rather nervous for such imposing birds, and are scarce visitors to gardens. Typically, groups of birds perch high amongst the branches of bare trees in winter, congregating in old nests as dusk falls.

Bulkier than Carrion Crows, Rooks have a pointed bill, pale face, high forehead and a shaggy 'trousered' look about the legs. Adults (**left**) are glossy black all over with a purple sheen, particularly on the head and wings. The base of the bill, chin and lores shows a large patch of whitish bare skin. The broad-based, tapering bill is silvery with a dark tip.

Youngsters (**right**) resemble Carrion Crows, but can be told by their size and the shape of the head and 'trousers'. The plumage is dark brown, the face feathered. The bill looks darker than the adults'.

The call is a distinctive 'kaah'.

135

Carrion Crow and Hooded Crow

Corvus corone/Corvus cornix Length: 44–51cm; wingspan: 84–100cm

Both species are common breeding residents within their range. The Carrion Crow is found countrywide except for the middle of Scotland northwards and the whole of Ireland, where it is replaced by the Hooded Crow. As with other members of the Crow family, both species are rare, nervous visitors to gardens, preferring farmland, hills, cliffs, moorland and woodland fringe.

Carrion and Hooded Crows share heavy black bills, round foreheads, flattish crowns and square tails.

The Carrion Crow (**below**) is entirely black, with a hint of purple gloss. It appears compact compared to the larger Rook. The heftier, rounder bill, 'shorts' rather than 'trousers' and squarer tail also help to differentiate it from the Rook.

Hooded Crows (**below**) are structurally identical to their cousins, but can be instantly told by their striking grey and black plumage - while they have a black head, breast, wings and tail, the rest of the plumage is battleship grey, which presents a striking image in flight. On a close view, note the fine black streaks on the underwing, upper breast, mantle and flanks.

Both Hooded and Carrion Crows are opportunistic feeders with a well balanced diet. They happily eat fruit and grain, insects, worms, small mammals, birds' eggs, nestlings and carrion.

The call of both species is a harsh 'kraa' or 'caw'.

Starling

Sturnus vulgaris Length: 19–22cm

The Starling is a common, widespread resident breeding species, and an autumn migrant – many hundreds of thousands of birds from northern and eastern Europe move westwards in the late summer and into the autumn. A bickering, noisy visitor to almost any garden anywhere, it is also found on farmland, in woodland and on cliffs.

The Starling is slightly smaller than the Song Thrush, with a slim, pointed bill and peaked forehead. It appears blackish, but close up, the head and entire underparts have a distinctive purple iridescent gloss, fading to bottle green on the flanks. The upperparts have fine flecks on the mantle, rump and wings. Male Starlings show fewer spots than females.

Juveniles are buff grey-brown all over, except for a whitish throat and greyish mottling on the underparts. When they begin moulting, their 'half and half' transitional plumage can look quite odd.

Starlings are amazing mimics when singing, making incredibly accurate imitations of many other species.

House Sparrow

Passer domesticus Length: 14–16cm

One of the most instantly recognisable garden birds, the House Sparrow is a common resident breeding species, seen across the country throughout the year. They also like hiding in bushes and scrub, or chirruping away on roof tops and buildings. The species suffered an unexplained decline in the late 1980s and early 1990s. From the millennium onwards, the decline halted and numbers seem to be stabilising, with hints of an increase in numbers.

House Sparrows are small, rounded birds. The distinctive feature of the male (**left, top**) is the head, a complex pattern of grey, black, brown and white. The upperparts are brown, with darker streaking; the underparts are washed grey. The winter male looks rather dull, with a paler crown and bib flecked with white.

Females (**left, bottom**) have a broad creamy patch behind the eye, contrasting with a tawny crown, black eyestripe, greyish cheeks and white chin. The wings show straw yellow 'tramlines', bordered by black and merging into golden browns, charcoal blacks and white.

Juveniles resemble females, but are generally brighter, owing to their fresh new feathers.

The call is a simple 'chirrip'.

Tree Sparrow

Passer montanus Length: 12.5–14cm

The Tree Sparrow is a fairly scarce breeding and migrant species, seen throughout the year across most of Britain with the exception of the south-west, much of Scotland and Ireland, and the whole of Wales. Favouring arable farmland, parkland and woodland fringe, Tree Sparrows are only occasional visitors to gardens.

Tree Sparrows are smaller and neater than House Sparrows, but share a round head, shortish wings and plump body. The plumage of both sexes is surprisingly bright. An adult has a chestnut cap, black bib and cheek patch, with a white neck collar. The mantle is warm brown with bold streaks. The unmarked rump is yellow-buff, with a dark, notched tail. Note the double white wing bar. The underparts are white with buff flanks.

Juveniles are a more subdued version of adults. The head pattern shows a greyish wash to the centre of the crown, merging into a duller chestnut cap. The face pattern is more subdued and the cheek spot not so defined.

The call is a quick, resonant 'tek-tek'.

Chaffinch

Fringilla coelebs Length: 14.5–16cm

The Chaffinch is a common breeding and migrant species. Migrants from the continent arrive from September onwards, sometimes in very large numbers. A regular visitor to gardens across Britain, Chaffinches are also found in woodland, around farmland and in parkland at any time of year.

The Chaffinch is a slim, long-tailed sparrow-sized bird. Males (**far right**) have a blue-grey nape and crown, contrasting with a black forehead and russet cheeks. Note the pale lime-green rump fading to a grey central tail. The rest of the tail is black with white outer feathers. The wings show a prominent white shoulder patch and wing bar.

Females (**near right**) look much plainer than males, lacking any bright tones. The upperparts are generally fawn, with darker head stripes and mantle. The rump patch is smaller and duller green than the male's. The pattern on the wings is similar, if a little duller.

Chaffinches have a bounding, undulating flight. Note the greenish rump and the wingbars.

The call is a soft 'chip' or 'tink'. The song is a scratchy mix of semi-melodious notes, strung together in a descending tune.

Brambling

Fringilla montifringilla Length: 14–15cm

Bramblings are relatively common passage migrants which arrive, countrywide, from September onwards, leaving us for their northern European breeding sites by late winter. Gardens on the edge of beech woodland have the best chance of luring Bramblings to feeders and, in some winters, they can be frequent garden visitors.

The Brambling is similar in size and shape to the Chaffinch; in fact the two species are often seen side by side. A winter male (**below left**) (identified as male by its blackish head, winter by its grey nape) has orange shoulders and breast, thin white wingbars and white rump patch. The mantle is scalloped black on a brown background colour. The underparts fade from orange on the breast to white on the belly, with dark flecks on the flanks.

Females (**below right**) have a drabber head pattern than many males and the mantle shows dark brown scalloping. The orange on the wings and breast is more subdued.

In flight, the white rump patch is immediately obvious.

The call is a nasal 'te-eup' (like a squeaky hinge). There is also a hard-sounding flight call, 'tyeck'.

139

Greenfinch

Carduelis chloris Length: 14–15cm

Perhaps the commonest finch to venture into gardens, the Greenfinch is a widespread breeding species, seen throughout the year across the country. It is also found in parkland fringe, hedgerows and bushes.

The Greenfinch is a heavy bird with a stout head, body and bill. Males (**below left**) are generally green, with areas of yellow, black and grey plumage. Note the paler yellow rump, and yellow tail sides with black tips. The wings show a strong yellow leading edge.

Females (**below right**) are duller than males. The head is dark with a paler supercilium and moustachial stripes. The upperparts are dullish brown-green, with indistinct streaks on the mantle. The underparts are brownish-green, fading to greyish-white on the undertail.

Juveniles (**below**) resemble females, but their upperparts are more streaked and are generally browner above and yellowish-grey below. The underparts are quite heavily streaked.

In flight, note the greenish rump, bright yellow tail flashes and forked black tail. The yellow blazes on the wing are also prominent, being slightly brighter on the male.

The call is a hard 'jupp' or nasal 'tszwee'. The song is a strident twittering, delivered in a delightful butterfly display.

Siskin

Carduelis spinus Length: 11.5–12.5cm

A locally common breeding species, the Siskin is easiest to see in winter when it is a visitor to an increasing number of gardens. It breeds in conifer plantations, alder and birch woods across East Anglia, southwest England, and the west coast of Ireland and Scotland. As autumn turns to winter, migrants arrive from the continent and head for gardens across the country.

The Siskin is a small, neat finch, with a smallish head and a short forked tail. Males (**below, top**) are strikingly green and yellow, with a black forehead and bib. Note the black and yellow wings. The underparts show fine black streaks on a white belly.

Females (**right, middle**) are duller than males. The upperparts are lime green with fine dark streaks. The yellow on the wings is paler than on the male, and the underparts are whiter, with more streaking. Juveniles (**right, bottom**) show a distinctive pattern on the breast and flanks – grey smudges contrasting with clean whitish underparts. The upperpart colouration looks fresher too.

The forked tail and longish wings of the Siskin are easily seen in flight. The black and yellow on the male's upperwing is particularly obvious, but note also the yellow tail sides.

The call is a soft 'telu' or 'tl-leh'. The song is a rather scratchy affair, full of twitters and trills. Siskins are adept mimics too.

141

Lesser and Common Redpoll

Carduelis cabaret/Carduelis flammea Length: 11.5 – 14cm

The Lesser Redpoll is an increasingly scarce breeding species, and migrant, found in woodland, parks and heathland throughout Britain. Common Redpolls breed in the forests of Scandinavia and Russia. Numbers found in Britain in the winter vary from year to year – some years they can be extremely scarce, while in a good year, dozens mix with flocks of Lesser Redpoll. Redpolls as a whole have various racial guises and this accounts for their erratic European distribution. Both species are more than happy venturing into gardens.

Lesser Redpolls are round little birds with stubby bills, notched tails and gregarious habits. Common Redpolls (often still called Mealy Redpolls) (**left**) are larger, paler birds. The head and upperparts appear frosty grey-brown, the wing bars and rump are whiter and the underparts are paler.

Male Lesser Redpolls (**below right**) show a small red cap on the forehead, black lores and chin contrasting with buff cheeks and darker upperparts. The rump is generally pale pink and the wings dark brown with pale cream wingbars. The white underparts show a pinky-red wash on the breast and the flanks are streaked black.

Female Lesser Redpolls (**centre**) differ from males in showing only a small amount of red on the cap and very little, if any, on the breast. Also, the rump tends to be browner. This applies to Common Redpolls too.

The call is a buzzing 'jeet-jeet-jeet' uttered in flight.

Bullfinch

Pyrrhula pyrrhula Length: 14–15cm

Bullfinches are relatively common breeding birds seen throughout the year. They are also a very scarce migrant from northern Europe. For such dazzling birds, they are shy and retiring, slipping unobtrusively in and out of larger gardens. They can also be seen in scrubby hedgerows, woods, plantations and orchards across the country.

Bullfinches are very round and compact with a decidedly neckless look, a chunky deep-based bill, rounded wings and a square tail. The male (**below left**) has a glossy black head, contrasting with the grey mantle. Note also the white wing bar. The underparts from the face to the belly are rich carmine red. The female's underparts are fawn with a white vent and undertail (**below right**). The black cap, grey mantle of the male or brown mantle of the female, black and white wings, white rump patch and black tail are all distinguishing features of Bullfinches in flight.

Juveniles (**below**) have plain brown heads and upperparts, lacking the black cap and grey mantle.

The call is a soft, low-pitched, rather mournful 'pi-yuh'.

Hawfinch

Coccothraustes coccothraustes Length: 16–17cm

A locally uncommon breeding species and a rare migrant, Hawfinches can be seen year-round but are best looked for in late winter. Breeding in mixed and deciduous parkland, they favour anywhere with hornbeam or oak and are a very scarce visitor to larger gardens. They have a rather patchy distribution across England, Wales and southern Scotland.

The Hawfinch is a large, sturdy bird with distinctive plumage, a triangular bill, a large head and short tail. The male's head (**below, front**) is russet, with black feathering around the bill base and chin. The grey collar merges into a rich mahogany mantle, russet rump and central tail. Close up, the remarkable lyre-shaped tertials of the male can be seen. The female (**below, back**) is duller and more grey-brown than the male, particularly on the head, rump and underparts. There is less black around the face, and the secondaries and primaries are greyish, not black.

In flight, note the short, square tail and big head and bill. The broad white wing bar on the inner wing, white shafts on the flight feathers and broad white tail band are all obvious.

The call is a metallic 'zik-zik'.

Chapter Eight
Mammals

There are 4,000 species belonging to the class *Mammalia* worldwide, 200 of them within Europe. Warm-blooded vertebrates that breathe with lungs, mammals have a regulated body temperature, maintained by body hair, and give birth to live young, which are suckled by the female.

The mammals covered here fall into five different orders. The Western Hedgehog, the Mole and the two Shrew species mentioned come within the most primitive order, *Insectivora*. These are generally small, ground-living creatures, which feed mainly on insects. *Chiroptera*, the order of bats, also have furry bodies and feed on insects, but also have membranous wings with which they fly. The predatory mammals of the order *Carnivora* – the Red Fox (from the family *Canidae*) and the Weasel, Stoat, Pine Marten and Badger (from the *Mustelidae* family) – are characterized by enlarged canine teeth, which cut and tear through flesh. Perhaps the most unusual visitors to a garden are the Roe Deer and the Muntjac. Both species are even-toed ungulates (hoofed animals), belonging to the order *Artiodactyla*. Mice, voles, rats and squirrels belong to the order *Rodentia*, the primary characteristics of which are two pairs of front teeth, which grow continuously as they are constantly worn down by use. The final order of mammals here is *Lagomorpha*, whose sole representative is the Rabbit. All lagomorphs are herbivores, with a unique set of double front incisors.

Western Hedgehog

Erinaceus europaeus Length: 21–34cm (including tail)

The Hedgehog (or Western Hedgehog to be more accurate) is an unmistakable garden visitor, widespread throughout Britain and Ireland, as well as western Europe to Russia, and seen in both rural and urban gardens. Scrub, areas of woodland and hedgerows are favourite habitats, along with meadows and pasture.

The Hedgehog is quite variable in size, and is covered in around 6,000 spines (each 2–3cm long) from the top of the head to its rear. The gentle, pointed face and the entire underside are covered in coarse pale buff hair. Note the teddy bear button black eyes and the slightly upturned snout. Hedgehogs have small ears, which are often partially hidden by hair. Their legs are surprisingly long and each foot has five toes.

The Hedgehog is generally nocturnal, but is also seen during daylight hours. Adept at climbing banks and walls, Hedgehogs are also competent swimmers. Its diet is varied, consisting mainly of invertebrates found at ground level (worms, beetles, slugs, caterpillars etc.), but also including frogs, lizards, birds' eggs, nestlings, plants and even fish. When feeding, Hedgehogs snort and snuffle – some of the few noises they make.

Largely solitary, Hedgehogs spend night after night feeding on their own and do not form territories. Individuals only really make contact with each other after hibernation (which lasts from October to March or April), when mating takes place.

The female builds a nest of grass and leaves in which to give birth (a similar nest is built for hibernation). She gives birth to between four and six blind, spineless young, generally just once a year. The spines develop very quickly, and after three to four weeks, the young leave the nest. The male has no part in rearing the young.

Hedgehogs live for around three to four years (sometimes double that) and in this country, they have very few predators. They fall prey to Red Foxes and Badgers, and they are susceptible to poisoning, especially from slug pellets. Most casualties, however, are road-kills. When danger is near, they adopt their famous curled-in-a-ball posture.

146

Common Shrew

Sorex araneus Length: 5.2–8.7cm (up to 13cm including tail)

If your garden backs on to woodland or has an area of thick grass, you may be lucky enough to see a Common Shrew. These creatures are found throughout Britain (but are absent from Ireland), and are happy anywhere that has enough ground cover, favouring areas of scrubby woodland, hedgerows and rough areas of grassland. They are quite aggressive outside the breeding season, and will defend their territory ferociously.

The adult has a distinctive three-toned coat. The head and back are dark brown, contrasting with paler honey brown on the flanks and greyish-white on the underparts. Note the impressive snout – a longer, more whiskery affair than you would see on a mouse. A young Common Shrew is paler and has a hairy tail, compared to a young mouse.

Common Shrews are almost constantly active. They feed for short spells, during both day and night, before returning to their nests to rest. After a short nap, a quick snack on food saved from previous outings and a grooming session, the shrew is out hunting once more. To survive, it has to eat almost its entire bodyweight in beetles, snails, spiders and the like every day.

Common Shrews breed any time between March and September. A pair may have as many as five litters in a season, but the mortality rate of young is very high, with only around 30 per cent surviving to the following year. The gestation period is between two to three weeks and the female gives birth to up to 10 naked, blind young. Within a further three weeks, they are fully weaned.

Pygmy Shrew

Sorex minutus Length 4–6.5cm (up to 11cm including tail)

The Pygmy Shrew shares many of the same habitats as the Common Shrew, favouring rough pasture and grassland. Found throughout Britain, the Pygmy Shrew is a secretive garden visitor, and can sometimes be seen close to the warmth of a garden compost heap. It is an extremely agile little mammal, quite happy to clamber through vegetation as it forages for food.

An adult is similar in size to a young Common Shrew, but can be told apart by its grey-brown head and upper body, contrasting with its off-white chin and belly. Despite being smaller than the Common Shrew, the Pygmy Shrew has a rather beefy appearance with a broader and more rounded head and a thicker, hairier tail. If your cat presents you with a dead shrew, take a look at the teeth – those of a Pygmy Shrew will appear reddish.

The Pygmy Shrew has to feed voraciously to survive, eating well over its own body weight in small snails, slugs and insects every day. As with the Common Shrew, the mortality rate is high; many Pygmy Shrews are thought to die within their first couple of months. Even after that, their life cycle ends within about a year, during which they must be wary of cats and wild predators such as Stoats, Tawny Owls and Barn Owls.

Pygmy Shrews breed between April and October, with the emphasis on June and August. They do not dig their own burrows but borrow them from other small mammals. The gestation period is just over three weeks, after which four to seven blind, naked young are born. Three weeks later, the fully weaned young shrews are ready to move on.

Common Mole

Talpa europea Length: 13–20cm (including tail)

The Common Mole is a delightful animal if you are someone who loves sharing space with the wildlife around you. To some gardeners, however, it is a pest whose molehills are intolerable. The Mole can be found across almost the whole of Britain, but is absent from Ireland. Primarily a woodland species, this highly adaptable creature is found in fields, gardens and parks, particularly in areas where the soil is well drained and easy to dig.

When seen above ground, the Mole is unmistakable, with its soft, grey-black fur, rounded cylindrical body, huge earth-moving front paws and pink nose. A closer look reveals tiny eyes, covered almost entirely by fur, more normal-sized back paws and a neat little tail, often held erect.

The molehills that are dotted around gardens, pasture, woodlands and meadows across the country are indicators of a mass of activity below ground. Moles spend their lives within a complex series of burrows, created by their large, spade-like front paws. These burrows can be up to 1m in depth and over 200m in length, and consist of semi-permanent tunnels, nesting and sleeping chambers. The nesting and sleeping chambers are lined, usually with grass, moss and leaves.

The Mole has a reasonable sense of hearing and a terrific sense of smell, which enables it to seek out earthworms (which make up a massive percentage of its diet), insect larvae and occasional amphibians. It keeps tabs on where other Moles are via a series of squeaks, although they seldom fight.

Mating generally occurs between April and May, when males abandon their territory in search of a mate. The naked young are born around a month later. The fur develops in around two weeks, the eyes open after the third week, and the young leave the nest chamber after roughly five weeks.

The Mole's average life span is two to three years, but some live to be as old as seven. As well as irate gardeners, its enemies are cats and dogs, owls, Stoats, Red Foxes and Buzzards. Drought and floods can also be life-threatening.

Common Pipistrelle

Pipistrellus pipistrellus Length: 3.5–5cm (up to 8.5cm including tail); wingspan 27–30cm

The Common Pipistrelle is the most common British bat, found throughout Britain and Ireland. It often flies over gardens or under streetlights after dark, and is also frequently seen along hedgerows, woodland fringes and near water. Common Pipistrelles are one of several species that will roost and breed in houses.

Often all that will be seen of a Common Pipistrelle is a small dark shape, flitting by at dusk during the spring, summer and autumn, and on warmer winter days. Common Pipistrelles fly with a quick, rather jerky flight that makes the bat's small size and wings very obvious. Some people are able to hear them call, but they are generally beyond our hearing range.

If you see a Common Pipistrelle in other circumstances (they can be grounded on occasion), their tiny size will become apparent. With a body length of no more than a few centimetres, it will fit comfortably in the palm of a hand. Common Pipistrelles have a soft reddish-brown coat, varying to darker or paler shades of brown. They have short black ears and a black snub nose (typical of all bat species). The wing membrane is also black.

Common Pipistrelles head out to feed just after sunset, weather depending, and hunting forays may last for several hours. They feed on insects - midges, small moths, lacewings - and studies have shown that up to 3,000 can be eaten in a single night.

Breeding takes place between August and November. A male has up to 10 females to breed with, but fertilization is put on hold until the following April. The female then gives birth to one youngster in June or July. The birthing process is often synchronized within a colony, with around 80-100 bats present. The species has an average life span of around four years.

Noctule

Nyctalus noctula Length: 6–8.5cm (up to 14cm including tail); wingspan 32–40cm

At almost twice the body size and with a significantly broader wingspan, the Noctule is much larger than the Common Pipistrelle. It is often seen before sunset, sometimes in quite bright conditions. Although they have been in decline (along with so many other bat species) Noctules are still fairly widespread within their range in this country. They are present throughout England and Wales, but absent from Scotland and Ireland. Prime habitats are deciduous woodlands, parkland and gardens with, or close to, mature trees.

The Noctule's flight action is relatively smooth and, as with other larger species such as Daubenton's Bat and Natterer's Bat, it is capable of nifty twists and turns whilst on the wing. The flight silhouette should reveal a wedge-shaped tail and slender, angled wings.

If you see a Noctule in good light, you may be able to see its golden-brown fur, which is slightly darker on the back. When the Noctule moults during the late summer and early autumn, this golden hue becomes darker and duller. The dark brown to black wings are obvious, but the large, rounded ears and snub nose may be very hard to see.

The Noctule feeds anytime from before sunset to after sunrise, but typically feeds for an hour or so around dusk. Typically, it then makes one night-time outing, and another around dawn that can last well into daylight. It is seen throughout the year, as long as suitable winter conditions prevail, usually skimming over treetops or flying over open pasture in search of large insects, moths and flies.

The breeding cycle begins in August to October when mating takes place but, as with Pipistrelles, the female retains the sperm and fertilizes her eggs the following spring. The young bats are born in June and July. Females reach sexual maturity after just three months, the males around a year later.

Brown Long-eared Bat

Plecotus auritus Length: 3.5–5.5cm; wingspan 23–28.5cm

The Brown Long-eared Bat (also known as the Common Long-eared Bat) is exactly as the name suggests – brown, with long, very obvious ears, which can be up to 4cm long. The species is widespread and relatively common across Britain and Ireland.

Although almost identical in wingspan and body length to the Common Pipistrelle, its big, broad ears make the Brown Long-eared Bat appear larger. This species shares with the Pipistrelle a fast, erratic flight pattern, but the long ears are readily apparent even in silhouette. In good light, the fluffy fur on the body can be seen to be grey-brown above and markedly paler below. The wing and tail membranes are much paler than those of either the Pipistrelle or the Noctule, but like them, this bat has a squashed-up face and a small snub nose.

Brown Long-eared Bats embark on their first feeding flight well after sunset, and it lasts around an hour. They make several short flights during darkness, returning to their roost before dawn. On the wing, they feed on a variety of insects, including moths, beetles and flies.

Breeding generally takes place in autumn, before hibernation. As with the previous species, the female is able to suspend fertilization until spring, and give birth to a single offspring in June or July of the following year.

Favoured areas are sheltered spots near deciduous or coniferous woodland, parkland and well-wooded gardens. They will make good use of bat boxes and the attics of houses during the summer months, and will also utilize buildings for their winter roost sites.

Brown Long-eared Bats live for an average of four to five years, but have been known to reach an amazing 22 years old.

Red Fox

Vulpes vulpes Length 85–120cm (including brush)

Appearing larger and leaner than a domestic cat, with a long bushy tail, a pointed nose and ears, rather long legs and a reddish-brown and white coat, this lovely animal is unmistakable. The Red Fox is now familiar both in the countryside and the city, and can be seen across the whole of Britain and Ireland in almost any location, providing there is sufficient cover. It is a frequent garden visitor, especially in towns and cities, and, despite persecution from humans, it continues to thrive. Indeed, it is one of the most widespread carnivores in the world.

The Red Fox is largely russet brown with white on the muzzle and face, along the belly and the inside of the legs. The pointed ears are tipped black and the front of the legs and the paws are also blackish. The long bushy tail (the brush) is russet with an obvious white tip. The male (the dog) is usually some 15-20 per cent larger than the female (the vixen).

Primarily nocturnal, the Red Fox is also a familiar sight before dusk, after dawn and, in the breeding season, during daylight hours if hungry cubs need food. Its diet is wide and varied. Rabbits, hares, ground-nesting birds, voles, mice and rats make up the meat content (along with domestic chickens on a rural menu), while the urban Red Fox scavenges on carrion and human refuse. During the summer and autumn, it also feeds on berries and fruit, while the cubs find important nourishment from earthworms.

In midwinter (usually December or January) the distinctive, rather spooky yelping triple note 'whoow-whoow-whoow' echoes around towns and cities as well as open countryside, and heralds the start of the mating season. The dog and vixen spend several days together at this time, with the dog guarding the vixen from other interested parties. After a period of around eight weeks, the vixen gives birth to four to five cubs in her earth. Blind and deaf at birth, the cubs have dark brown fur, which takes on a red tinge after around a month. Within six weeks the cubs are catching worms and insects for food and are fully weaned from the vixen after seven weeks. They mature to adulthood within around six months.

Stoat

Mustela erminea Length 32–44cm (including tail)

The Stoat is a lithe, slender carnivore, larger and longer bodied than the similar looking Weasel, with a decidedly longer tail. It is common and widespread across Britain and Ireland.

The Stoat has a small, rounded head and a longer neck than the Weasel (when alarmed, the Stoat extends its neck, making it particularly noticeable). The fur on the upperside is warm brown in colour, extending from the nose to the tail. The tail tip is black, the underside creamy yellow. The front legs are clearly shorter than the back, and the leg fur tends to be whiter than on the belly. The male is larger than the female.

In the northern areas of Britain, and elsewhere in northern Europe, the winter moult (one of two during the year) sees a marked transformation in the Stoat's appearance – the brown fur is exchanged for the white ermine so relished by furriers. The tip of the tail remains black, and the moult back to brown fur takes place in March and April. In some areas of Britain, the Stoat undergoes only a partial moult to ermine; in southern Britain, the winter coat remains brown.

Stoats are happy hunting by night or day. Favoured prey includes voles, mice, ground-nesting birds, Hares and Rabbits. (The blood-curdling squeals of a Rabbit when killed with a sharp bite to the back of the neck are unforgettable.) Stoats bound along with a decidedly springy gait, stopping in an instant, looking around at anything that catches their attention, and then bounding off again. They are also competent climbers and swimmers.

Mating takes place during summer, but the implantation and gestation is delayed for many months (generally around nine). The female eventually gives birth to as many as 12 young in April and May – almost a year after mating occurred. The young spend six weeks blind, prior to weaning, and are independent after about three months.

The Stoat is widespread and common across the country, in a wide variety of habitats providing there is cover - large gardens and farmland to moorland and mountains. It makes its den in hollow trees, rock crevices and burrows (often those of its victims).

Weasel

Mustela nivalis Length 19–29cm (including tail)

The Weasel is relatively common and widespread across the whole of Britain but is absent from Ireland. It can occur in almost any habitat that provides enough cover and food, from large gardens to woodlands, dune systems, grassland and moorland. The Weasel is noticeably smaller than the similar Stoat, with a slightly broader head, a shorter neck, a more rounded and shorter body and a much shorter tail. However, it is just as tenacious when out hunting.

Weasels are wholly reddish brown above (including the whole of the short tail), with white underparts rather the Stoat's creamy yellow. The broader head and more prominent snout are quite different from the Stoat's, and the front legs are in better proportion to the body.

Weasels are active by day and by night, alternating periods of activity with a few hours' rest. Their bounding gait is similar to the Stoat's, but their shorter legs mean they hurry and scurry more. Favourite foods are voles and mice, but also young Rabbits, birds and birds' eggs in spring and summer.

The breeding season falls anywhere between February and August. The gestation period is just over a month and the female Weasel gives birth to four to six kits. There is usually just one brood, but two in a year is not unusual. The kits are weaned at around eight weeks and become totally independent by three months of age. Some females, if food is abundant, are able to breed three to four months after birth. In the wild, the Weasel's life expectancy is around a year.

Pine Marten

Martes martes Length 66–80cm (including tail)

It is hard to imagine any garden visitor that is as charismatic and beautiful as the Pine Marten. With a very restricted range across Britain and Ireland, they are prone to suffer when habitats are disturbed and have been persecuted in the past. The best place to see them is in north-western areas of Scotland, where they frequently come to gardens (and have apparently developed a taste for marmalade sandwiches!). There are also populations in the Lake District, north Wales, isolated parts of Yorkshire and several areas across Ireland.

Pine Martens have a flattened look to the head, with a soft expression created by the combination of bright black eyes, neat, rounded snout and neatly proportioned ears. They have quite a long neck, longish legs and a long, bushy tail. The fur is rich brown, darker on the paws, contrasting with a creamy yellow throat. Even the feet are furred. Similar in size to a domestic cat, they are larger than Stoats, and larger, broader and more muscular than the Polecat or domestic ferret.

This species is almost entirely nocturnal, although a mother and her cubs will move during the daytime through the summer and early autumn. The diet is seasonal, with voles, birds, frogs and mice eaten in summer, and as autumn approaches, the Pine Marten also seeks out berries, mushrooms, beetles and honey.

Mating takes place between July and August, with implantation usually following in January or February. After a month's gestation, the cubs are born in March or April. Den locations vary, but are often old squirrel dreys or hollow trees. The litter size averages three cubs, which remain within the family group for six to seven months before becoming independent. As long as they survive the first winter, Pine Martens can expect to live for around five years.

European Badger

Meles meles Length 78–105cm (including tail)

With its heavy build, short legs, short tail, long snout and distinctive colouration and markings, the Badger is unlike any other garden visitor. Its upper body is grey, contrasting with its black underside, black legs and stripy black and white face. The tail is usually tipped white, as are the largely greyish ears. Although found across the whole of Britain and Ireland, it is a secretive animal.

The male (boar) has a broader, more domed head than the female (sow), who has a narrower, flatter profile. The sow also has a bushier tail. The strong claws (especially heavy-duty on the front paws) are shared by both sexes and are excellent digging machines for both excavating a sett or foraging for food.

A Badger sett is a communal affair. Generally, it comprises a number of entrances spread across an area of many metres, all of which lead into a series of underground tunnels that run for up to 20–30m below ground. Within the tunnel system are sleeping chambers and underground latrines, the latter situated well away from the main areas of the sett. Badgers gather bedding material such as straw, grass or leaves with their forelegs and chin, dragging it backwards towards the sett. Some setts can be active for many Badger generations; they can stretch for some distance and become remarkably complex in their layout. Badgers favour areas of deciduous woodland for setts, but will also settle in large overgrown areas, field edges, cliffs, buildings and even caves. In some areas, the Badger is a regular night-time garden visitor, enticed by milk or pet food.

Badgers mate throughout the year, but usually between February and May. Delayed implantation is common, and most cubs are born, in the sett, early the following year. A brood of up to five cubs is born and takes around three months to be weaned. After a further three months they are independent of their mother, but usually stay within the sett until after their first winter.

Like many other carnivores, the Badger has a well-balanced diet. It is an opportunistic feeder, eating earthworms, beetles, nests of both bees and wasps, birds' eggs, voles, mice, Rabbits and carrion. They will also turn to fruit such as raspberries, or oats, wheat and nuts, especially during the late summer and early autumn.

Muntjac

Muntiacus reevesi Length 90–120cm (including tail)

The Muntjac (more correctly known as Reeves' Muntjac) is a small russet-coloured deer, barely the size of a Labrador dog, and is a very occasional visitor to large gardens. It was introduced to southern England in the 1900s, and is the smallest deer found in Europe, as well as being one of the most attractive.

The Muntjac is a shy, short-legged and rather rounded-looking animal. The body is mostly rich reddish-brown, with the exception of the buff underbelly and ginger tail. The underside of the tail is white, but this is only seen when the tail is lifted. In the autumn, the coat becomes very dark chestnut brown and the legs almost black. Both sexes share a characteristic black 'V' mark on the forehead. The male (the buck) has short, backward-pointing antlers (only a few centimetres long), and long incisors, that frequently protrude like miniature tusks.

Something of a loner, a Muntjac is active throughout the day, particularly around dawn and dusk. A browser and a grazer, it will eat grass, shrubs and herbs, along with some fruits and nuts. Its grazing habits have presented some problems in areas of ancient woodland.

There is no recognised breeding season in England. Bucks announce their intentions via a series of barks, repeated frequently over a minute or so. Once a pair has mated, the gestation period is around seven months. A single fawn is born, and the female (the doe) quickly comes back into season. The fawn is usually weaned after four to five months. As with other deer species, the Muntjac fawn is spotted for the first two to three months of its life, and lies quietly in vegetation, close to the female.

Muntjacs have increased across much of their range in southern England and have become more common with each passing year. They favour areas of dense woodland with plenty of thick undergrowth, but they will emerge from cover and feed quietly at the edge of large gardens and orchards. Sadly, the species is a victim of its own success – many die as road casualties, some are shot and predators, such as the Red Fox, take fawns. A Muntjac surviving all this can expect to live for up to 15 years.

Roe Deer

Capreolus capreolus Length 100–140cm (including tail)

The Roe Deer is found across much of Scotland and in a narrow strip running across the centre of England from East Anglia to the far south-west. They are absent from Ireland.

The Roe Deer is larger than the Muntjac, being taller with a long body and longer legs. Its face is also very different, appearing gentler and more attractive. The coat is reddish-brown (perhaps slightly richer in tone than the Muntjac's) with pale areas around the big, liquid, black eyes, the black nose and muzzle, the inside of the ears and the underside of the body. Roe bucks' antlers are longer than a Muntjac's and far more upright. Both sexes appear not to have a tail; the eye is drawn to a large 'powder puff' white patch on the deer's rear end. As winter approaches, Roe Deer moult to a much darker, greyer coat.

As with other deer species, the Roe Deer is shy, but perhaps less so than the Muntjac. Very occasionally, if you remain still and the deer doesn't smell you, it may approach to within a few yards. Both sexes are territorial throughout the year, but the territories of male and female overlap constantly.

Roe bucks take up their breeding territories from April to August. Rutting usually occurs through August, and, after mating, the implantation is delayed, usually until midwinter. Once the doe truly falls pregnant, the gestation period is around four to five months. The doe gives birth between late spring and early summer, often bearing twins, which are fully weaned after around three months. The fawns are dark with several rows of white spotting. They remain with their mother until the following year, being chased off just prior to the doe giving birth to a new fawn.

The Roe Deer feeds on only the tastiest morsels from the hedgerow or the ground. Brambles, grasses and young, emergent leaves on broad-leaved trees are top of the list during summertime, with beechmast, mushrooms and acorns joining the list in autumn. On moorland, where Roe Deer can be common, they feed on heather and bilberries.

Roe Deer will pay furtive visits to large gardens, especially those that border deciduous or coniferous woodland. In Scotland they are often seen on moorland fringes, bounding through the heather as they head for the safety of a nearby conifer plantation. As long as they avoid roads and hunters, Roe Deer live for between 10-14 years.

Grey Squirrel

Sciurus carolinensis Length 23–30cm (not including tail of 19–24cm)

A regular visitor to even modest-sized gardens, the Grey Squirrel is a relatively recent introduction to this country – brought over from America and released at a number of locations between the 1870s and 1920s. Since then, it has spread in huge numbers and the species is extremely common across much of England and Wales, and has spread into southern areas of Scotland. It has also been introduced into Ireland, where it is found in eastern and central areas. The Grey Squirrel is viewed as a pest in many areas and is seen as a major reason for the decline of the Red Squirrel, though this is open to debate.

The Grey Squirrel is larger and more robust than the Red Squirrel, with smaller ear tufts, a grey back and a grey tail. The head is warm brown in colour, merging into the grey upper body and contrasting with the white throat and belly. The longish tail is grey with a distinctive rusty tinge and a white tip. Occasionally, during the summer, some Grey Squirrels take on a reddish hue, but the size, build and lack of ear tufts help to identify it from the scarce Red.

Grey Squirrels nest in round, compact dreys consisting of twigs and leaves on the outside, and a soft bed of dry leaves and grass on the inside. They may use several dreys over the course of the year, where they both give birth and find shelter.

A male will venture towards a female, full of boisterous, noisy 'chattering'. After successful mating, the female will give birth, eight weeks later, to three or four blind and naked young. The baby squirrels are cared for solely by the female and, after four months, they are completely independent.

Grey Squirrels are active for several hours a day during spring and summer. This increases to all day in autumn, as the animals busily gather and store food for the winter months, hoarding acorns, nuts and beechmast just below ground or in tree hollows. During the short days between November and March, they are generally active for a short space of time in the morning.

Favoured foods include acorns, beechmast, nuts, fruits and flowers. Insects and birds' eggs are occasionally eaten too, and Grey Squirrels are often seen climbing up onto bird tables and bird feeders in search of an easy meal.

These are certainly not shy animals and they can become very tame and approachable, particularly in parks. Grey Squirrels can be seen in mixed and deciduous woodland as well as parks and gardens, and they can live for up to nine years.

Red Squirrel

Sciurus vulgaris Length 21–25cm (not including tail of 14–20cm)

Along with the Pine Marten, the Red Squirrel is one of the most attractive of all garden wildlife visitors. Because of its restricted range across the country, you will only see one if you live in the Highlands of Scotland, tiny pockets of Dorset, East Anglia, the Isle of Wight or in Ireland, where the species is widespread.

Often thought to have been forced out of its habitat by the Grey Squirrel, the Red has indeed suffered a decline since the arrival of its North American cousin. However, the loss of hazels and conifers is also to blame, as is disease. Red Squirrels are slightly smaller and less robust in appearance than Greys. The Red has a softer expression and more slender build, and its tufted ears are very noticeable. The whole coat is reddish-brown except for the white throat, chest and belly. The wispy tail is also foxy red in colour, though in summer this can become very pale, almost translucent. Very occasionally, brown or even black forms may be seen in this country, but these are much rarer here than on the continent.

Like the Grey Squirrel, the Red nests in a round, compact drey with an outer shell of twigs and an inside lined with grass and moss. It may use several dreys over the course of a year, both for raising young and sheltering during poor weather.

Mating depends on the availability of food – it generally takes place very early in the year, but if the cone crop is poor, Reds will mate in summer. Lots of 'chattering', tail flicking and chasing through the trees take place in the lead-up to mating. Afterwards, the female has a six to seven week gestation period before giving birth to between one and eight young (usually three to four). The young Red Squirrels are fully weaned after seven weeks and begin to venture out from the nest. After around three months, they are fully independent. The female is the sole carer.

Red Squirrels are most active in the morning, usually in the first hours after sunrise. After a rest, they re-emerge in the late afternoon or early evening (depending on the season) for another two or three hours. They are much less active during the late autumn and winter months, only emerging for a few hours in the morning.

Favourite foods include pine and spruce seeds, acorns (all of which they hoard away for later consumption), berries, fungus and occasional invertebrates and birds' eggs. Reds also nibble bark and soil for roughage.

Red Squirrels are found in huge blocks of conifer forest, but will also venture into gardens or parklands. In some parts of England they are still found in deciduous woodland. They tend to fall victim to birds of prey, Pine Martens, dogs, cats and traffic; but otherwise, they live for six to seven years.

Bank Vole

Clethrionomys glareolus Length 8–11cm (not including tail of 3–7cm)

The Bank Vole is a cute-looking mammal that is very common across Britain, but found in just one small area of western Ireland. It is rather hard to see because of its rather secretive habits.

Although shorter than the Field Vole, it does have a longer looking tail than its relative. It also has redder fur, a more rounded face and smaller eyes. Its upperparts are warm chestnut brown, contrasting with a grey-buff throat, chest and belly. The rather short legs and feet are fleshy pink in colour, as is the tail.

The Bank Vole is active by day and night, though it is more nocturnal in summer. It is most active around dawn and dusk, searching for fruits and berries, seeds, worms, insects, fungi, snails and, in winter, dead leaves.

The breeding season is generally between April and September, but can occur throughout the year. A female makes her nest in woodland or grassland, with a defined entrance and exit. A woodland nest is lined with moss and feathers, a grassland one with grass and moss. The gestation period is around three weeks, and the female gives birth to between three to five young. The baby voles open their eyes after 10–12 days, and within just three weeks, the young are fully independent. A female Bank Vole may have up to five litters in a year.

Bank Voles are found in both deciduous and coniferous woodland, providing there is a mass of undergrowth. They are also seen along banks and hedges and, of course, in gardens.

Several species of owl, Weasels, Stoats and Red Foxes all hunt the Bank Vole, with the result that its lifespan is rarely more than 18 months.

Field Vole

Microtus agrestis Length 9.5–13.5cm (not including tail of 2.5–4.5cm)

The Field Vole, like the Bank Vole, is a very attractive mammal – perhaps even cuter. The species is common across the whole of Britain, but is absent from Ireland.

The Field Vole is around 3cm longer than the Bank Vole, with a more yellowy, golden coat, a shorter tail and a slightly more elongated face. The underparts tend to be whiter, too.

The Field Vole is active by day and night, but prefers a more nocturnal lifestyle (particularly in summer) and is especially active at twilight. It eats herbaceous plants and many grasses, along with seeds and leaves. It uses a (sometimes large) underground burrow system, connected to mini-pathways on the surface.

Field Voles mainly breed between April and September, although young may be seen year round. The female constructs a nest of grass at the bottom of a large tussock. With a gestation period of just under three weeks, the female soon gives birth to between four and six young, which are weaned after a fortnight. A week later and the young Field Voles are independent of their mother.

The Field Vole favours areas of damp grassland – meadows, fields, rough areas of gardens – but can also be found near young plantations, open woodland or even mountain and dune areas, providing there is sufficient scrubby cover. As with the Bank Vole, this species tends to fall victim to owls and predatory mammals such as the Stoat and Red Fox. Very few Field Voles live beyond 18 months.

Wood Mouse

Apodemus sylvaticus Length 8–10cm (not including tail of 7–11.5cm)

The agile Wood Mouse is distinguished from other mouse species by its relatively large ears and eyes, and its large hind feet. It also has a very long tail, which can be longer than the mouse itself. Wood Mice are found across the whole of Britain and Ireland. Although woodland is their favoured habitat, they are also seen in gardens, patches of scrub and arable land, and at coastal sites too.

The species is largely warm orangey-brown above, with greyish-white on the muzzle, throat and belly. The flanks (and occasionally the neck) are yellowish-brown. The feet are very pale pink, as is the long, relatively thick tail. The large, rounded ears are fleshy pink, the big eyes black.

Wood Mice are generally seen only at night, although they may appear for an hour or two after dawn and before dusk during the summer. In winter, they are at their most active around sunrise and sunset. They eat a wide variety of foods, depending on availability – nuts and berries (especially blackberries), fungi, seeds (particularly oats and wheat), fresh buds and galls. They will also eat snails and earthworms. Adult Wood Mice may share communal nests in winter, and males often gather in self-made burrows during the summer.

The breeding season starts in March and lasts until October, with the late summer months of July and August seeing the cycle reach a peak. The female rears her young in a nest of grass, leaves and moss. After a gestation period of around three weeks, she gives birth to between two to nine young. Born naked and blind, the young Wood Mice develop quickly, and are fully weaned after three weeks. A female usually has one or two litters a year, but can have as many as four.

The Wood Mouse frequently falls victim to birds of prey, owls, cats, Weasels, Stoats and Badgers. Its life span is around 18–21 months.

Yellow-necked Mouse

Apodemus flavicollis Length 8.5–13cm (not including tail of 9–13.5cm)

The Yellow-necked Mouse is common across its range, south of a line running across the country from the top of Wales to East Anglia. Absent from northern England, Scotland and Ireland, the Yellow-necked Mouse is found in well-vegetated gardens, deciduous woodland, hedgerows and field edges.

The Yellow-necked Mouse is quite similar to the Wood Mouse in appearance, but is larger and brighter with more reddish fur on the upperside. The underside is whiter than that of a Wood Mouse and there is a notable demarcation between the upper- and underside. A distinctive yellow 'bib' can be seen on the Yellow-necked Mouse's throat, far larger than anything seen on a Wood Mouse. The ears are similar in shape and size, but the eyes are less bulbous. The tail tends to be shorter than the body whereas, for the Wood Mouse, this is generally the reverse.

A Yellow-necked Mouse becomes active at night, and has just one period of activity. During this time the mouse seeks out seeds, fruits, buds and also some moth and butterfly caterpillars and pupae. It will also eat birds' eggs.

Yellow-necked Mice breed between April and October, although they have been known to breed year round. The female builds a nest of leaves and grass, normally under a tree stump or in other mammals' burrows. She gives birth to around five young after a fortnight's gestation period. After just three weeks, the young mice are independent of their mother. The Yellow-necked Mouse has three litters over a year.

A frequent victim of predatory birds and mammals, this species has a life expectancy of around two to three years.

Brown Rat

Rattus norvegicus Length 13–29cm (not including tail of 10–20cm)

Extremely common across the whole of Britain and Ireland, the Brown Rat is probably the most disliked species within these pages – even more so than midges, wasps and spiders. But despite this lack of public affection, it is actually a rather attractive, shy creature that will do its best to keep out of the way of humans.

Brown Rats are quite bulky animals, with a neatly proportioned head, ears, eyes and body. They are covered in thick brown to grey-brown fur, paler on the underside and snout. The slightly protruding ears are pinkish, as is the long tail, while the eyes are deep brown.

The species is generally found in areas close to human habitation or areas where there are rich, easy pickings, such as rubbish dumps, farmyards and sewage systems. However, some Brown Rat populations live quite happily away from houses and factories, in crops and fields. Brown Rats have a hugely varied diet, from berries, cereals and seeds to birds' eggs, slugs and snails, along with scraps of meat and bones. They are active mainly at night or around dawn and dusk, but can be seen scurrying away from sight during daylight hours.

Brown Rats breed throughout the year. The female gives birth to around seven to eight young per litter, sometimes as many as 15, sometimes as few as just one. The gestation period is about three weeks. The young rats reach independence after two months and are themselves able to breed around a month later. Each female has four to five litters a year (if conditions prevail).

The species is very common in urban areas across much of the modern world. It is a carrier of disease and is dealt with by pest control workers across the whole of its range, but it also falls prey to owls and larger birds of prey, Red Foxes and cats. Most Brown Rats live for around 18 months in the wild, although some may live for up to three years.

House Mouse

Mus musculus Length 7–10cm (not including tail of 6.5–10cm)

The House Mouse is clearly different in appearance to the Wood Mouse or the Yellow-necked Mouse. It has a notably 'sharper' snout and less prominent eyes and ears than the Wood Mouse, to which it is closest in size. Originally found on the steppes of Central Asia, it has spread across Europe following grain cargoes, and to the Americas on the ships of explorers. The House Mouse is common across the whole of Britain and Ireland.

This mouse is generally grey-brown all over, except for a paler, white underside, but this is far less obvious than on either the Wood or the Yellow-necked Mouse. The eyes are small and black, lacking the strange bulging look of the Wood Mouse, and the ears are much less obvious. The legs and feet are fleshy pink, and the tail (almost as long as the mouse itself) is prominently ringed and scaly.

House Mice are largely nocturnal, feeding and drinking in the dark and resting during daylight hours. They love seeds (especially grain), but also eat invertebrates, roots and fungi. A House Mouse will spend some time skilfully de-husking grain in its front paws before eating it. They will also nibble at bars of soap in the house, along with candles, glue and plaster.

Providing there is a plentiful supply of food, House Mice breed throughout the year, particularly if nesting indoors. In more countrified surroundings, their mating activity peaks in late spring. The female makes a large nest of grass or any available material, including paper, cloth and sacking. After a three-week gestation period, around six naked, blind mice are born. These youngsters develop fast and are fully furred a fortnight after birth, weaned less than a week later, and are independent in less than a month.

The House Mouse is found in houses, factories, shops and warehouses as well as grassland, farmland and gardens. The species is rather communal in its habits, sharing nest spaces, particularly in colder weather.

This species has to be wary of birds of prey, owls, cats, Stoats, Weasels and, of course, humans. Its excellent balance and ability to jump and swim enable it to access many locations, and it can do a great deal of damage to grain and other food stores; it can also gnaw through wiring and other building materials. This, and its tendency to contaminate foodstuffs with its droppings, makes the House Mouse one of the least popular creatures within this book.

Edible Dormouse

Glis glis Length 13–19cm (not including tail of 11–15cm)

The Edible Dormouse is the rarest mammal covered by this book, found in just a handful of places in England. Introduced to Britain in 1902, it has a very restricted range in this country, being found only in the extreme south-east and south-west counties of England, and parts of south and west Wales.

With teddy-bear button black eyes, a gentle expression, a plump body and long bushy tail, this extremely rare garden visitor is unmistakable – and gorgeous! The coat is greyish, often with a brown or yellowish tinge. There is a faint dark line along its back, and tiny dark 'spectacles' around the eyes. The underside tends to be whiter than the upperparts. The long bushy tail is grey, and the ears are furless, rounded and rather small.

The species is nocturnal. Foraging in the canopy, it seeks out nuts and berries, fungi, and the occasional insect or bird's egg. It hibernates (often communally) between October and April; prior to hibernating, it puts on huge amounts in weight and fat reserves, as during its period of dormancy, it will lose about 50 per cent of its body weight. Given its night-time habits, the Edible Dormouse is very hard to see, as it spends daylight hours tucked away in a hole in a tree, an old bird's nest or inside a nest box.

Mating takes place between June and August. After a gestation period of a month, the female gives birth to between two to nine young in a nest made of moss high in the canopy, in a natural hole or in the cleft of a branch near the tree trunk. Very little is known as to how long the young take to become weaned and independent.

The Edible Dormouse favours areas of mature deciduous woodland, but has also been known to venture into houses, gardens and orchards.

And why the 'Edible' Dormouse? During the hedonistic times of the Roman Empire, the Dormice would be fattened up for the table in urns or jars, known as 'gliraria'.

Rabbit

Oryctolagus cuniculus Length 34–50cm (not including tail of 4–8cm)

With its long, rounded ears, familiar lolloping gait, soft expression and brown fur, the Rabbit is many people's favourite garden visitor. The Rabbit is found in large numbers across the whole of Britain and Ireland. The population has returned to very healthy numbers since 1953, when the myxomatosis virus caused a huge crash. The disease kills fewer and fewer Rabbits as decades go by.

Always smaller than the more rakish, gingery-toned Brown Hare, the Rabbit is a very easy animal to identify. The fur is grey-brown all over except for a distinct rufous patch on the back of the neck, and an off-white belly. Its long ears are tipped with brown, its prominent eyes are chestnut and it has a gently rounded nose. There are ginger tones to the fore- and hindlegs, and the underside of the tail is white – unmistakeable when the animal is startled and running away.

Generally seen around dusk and dawn or at night, Rabbits are often seen in quiet spots during the day. Their diet is entirely vegetarian and includes crops, cereals and saplings, along with grasses, bulbs and occasionally bark.

Renowned for their breeding activity, young Rabbits appear year round, peaking between late winter and late summer. Females born early in the year are able to produce their own youngsters later that year. Following a gestation period of about a month, females give birth to around five to six young in nesting chambers within the burrow. The babies are naked and blind at birth, but, within a month, they are weaned and independent.

Interestingly, both male and female Rabbits show aggressive traits towards young; males will attack any baby Rabbits, and a female will attack those that she knows not to be her own.

Rabbits occur in almost any habitat, but favour areas of heathland, meadows, grassland, farmland and woodland, as well as larger gardens.

REFERENCES

Attracting Birds to Your Garden
Stephen Moss & David Cottridge
(New Holland, 1997)

Attracting Wildlife to Your Garden
M Chinery
(Harper Collins, 2004)

Collins Field Guide – Butterflies of Britain and Europe
T Tolman & R Lewington
(Harper Collins, 1997)

Collins Field Guide – Insects of Britain and Europe
M Chinery
(Harper Collins, 1993, 3rd edition)

Collins Field Guide – Mammals of Britain and Europe
D Macdonald & P Barrett
(Harper Collins, 1993)

Collins Field Guide – Reptiles and Amphibians of
Britain and Europe
E N Arnold, J A Burton, & D W Ovenden
(Harper Collins, 1978, reprinted 1992)

The Complete Garden Bird Book
Mark Golley
(New Holland, 1996)

Field Guide to Dragonflies and Damselflies
of Great Britain and Ireland
S Brooks & R Lewington
(British Wildlife Publishing, 1997)

Field Guide to the Moths of Great Britain
and Ireland
P Waring, M Townsend & R Lewington
(British Wildlife Publishing, 2003)

FURTHER READING – BOOKS

Beetles
K W Harde
(Blitz Editions, 1999)

Bill Oddie's Introduction to Birdwatching
(New Holland, 2002)

Birds of Europe
Lars Jonsson
(A&C Black, 1996)

Bumblebees
Oliver E Prys-Jones & Sarah A Corbet
(Richmond Publishing, 1991)

Chris Packham's Back Garden Nature Reserve
(New Holland and The Wildlife Trusts, 2001)

Collins Bird Guide
Lars Svensson, Peter Grant, Killian Mullarney,
Dan Zetterström
(HarperCollins, 2001)

Field Guide to the Caterpillars of Butterflies and Moths
in Britain and Europe
D Carter & B Hargreaves
(Collins, 1986)

Field Guide to Spiders of Britain and Northern Europe
Michael J Roberts
(Collins, 1995)

Hoverflies
Francis S. Gilbert
(Richmond Publishing, 1993)

Land Snails of the British Isles
A A Wardaugh
(Shire Publications, 2000)

Minibeast Magic – Kind Hearted Capture Techniques
for Invertebrates
R Oxford
(Yorkshire Wildlife Trust, 1999)

Plant Galls
Margaret Redfern & R R Askew
(Richmond Publishing, 1998)

The Secret Lives of Garden Birds
Dominic Couzens
(Helm, 2004)

The Ultimate Birdfeeder Handbook
John A. Burton
(New Holland, 2005)

FURTHER READING – MAGAZINES AND JOURNALS

All the Wildlife Trusts publish their own magazines and there is the national Wildlife Trusts magazine, Natural World. The other organisations on the page overleaf all have their publications, but there are two commercially produced magazines that have features of interest to the garden naturalist.

British Wildlife
Lower Barn, Rooks Farm
Rotherwick, Hook
Hampshire RG27 9BG
www.britishwildlife.com

BBC Wildlife Magazine
Origin Publishng Ltd.
14th Floor, Tower House
Fairfax Street
Bristol BS1 3BN
www.bbcwildlifemagazine.com

The Birdwatcher's Yearbook
Published annually by:
Buckingham Press
55 Thorpe Park Road
Peterborough PE3 6LJ
Tel: 01733 561 739
buck.press@btinternet.com

USEFUL ADDRESSES

Bat Conservation Trust
15 Cloisters House, 8 Battersea Park
Road, London SW8 4BG
Tel: 020 7627 2629
www.bats.org.uk

British Butterfly Conservation Society
Manor Yard, East Lulworth
Wareham, Dorset BH20 5QP
Tel: 01929 400 209
www.butterfly-conservation.org

British Dragonfly Society
Dr W. H. Wain, Secretary
The Haywain, Hollywater Road
Bordon, Hampshire GU35 OAD
www.dragonflysoc.org.uk

BTO (British Trust for Ornithology)
The Nunnery, Thetford
Norfolk IP24 2PU
Tel: 01842 750 050
www.bto.org

C.J Wildbird Foods Ltd
The Rea Farm, Upton Magna
Shrewsbury, Shropshire SY4 4UR
Tel: 01743 709 545
www.birdfood.co.uk

Conservation Foundation
1 Kensington Gore
London SW7 2AR
Tel: 020 7591 3111
www.conservationfoundation.co.uk

Countryside Council for Wales
Plas Penrhos, Ffordd Penrhos
Bangor, Gwynedd LL57 2LQ
Tel: 01248 370 444
www.ccw.gov.uk

Department of the Environment
for Northern Ireland
Countryside and Wildlife Branch
Calvert House, 23 Castle Place
Belfast BT1 1FY
Tel: 01232 230 560

English Nature
Northminster House
Peterborough PE1 1UA
Tel: 01733 455 000
www.englishnature.org.uk
(Legal aspects of wildlife &
conservation, photographic
licensing advice)

Ernest Charles
Copplestone Mill
Copplestone, Crediton
Devon EX17 2YZ
Tel: 0800 7316 770
www.ernest-charles.com
(Birdfood)

FROGlife
Mansion House
27/28 Market Place Halesworth,
Suffolk IP19 9AY
Tel: 01986 873 733

Garden Bird Supplies
Wem, Shrewsbury
Shropshire SY4 5BF
www.gardenbird.com

Jacobi Jayne & Co
Wealden Forest Park
Herne Common, Herne Bay
Kent CT6 7LQ
Tel: 01227 714 314
(Birdfood)

J E Haith Ltd,
65 Park Street, Cleethorpes
Lincolnshire DN35 7NF
Tel: 0800 298 7054
www.haiths.com
(Birdfood)

Mammal Society
15 Cloisters House
8 Battersea Road
London SW8 4BG
Tel: 020 7498 4358
www.mammal.org.uk

Rob Harvey Specialist Feeds
Kookaburra House, Gravel Hill Road
Holt Pound, Farnham
Surrey GU10 4LG
Tel: 01420 239 86
www.robharvey.com

RSPB
The Lodge, Sandy
Bedfordshire SG19 2DL
Tel: 01767 680 551
www.rspb.org.uk

Scottish Natural Heritage
12 Hope Terrace , Edinburgh EH9 2AF
Tel: 0131 447 4784
www.snh.org.uk

Tree Advice Trust
Alice Holt Lodge
Farnham, Surrey GU10 4LH
Tel: 01420 220 22
Tree Advice Helpline: 09065 161 147
www.treeadviceservice.org.uk

Urban Wildlife Partnership
The Kiln, Waterside, Mather Road
Newark, Nottinghamshire NG24 1WT
Tel: 0870 0367 711
www.wildlifetrusts.org.uk

Watkins and Doncaster the Naturalists
PO Box 5, Cranbrook
Kent TN18 5EZ
Tel: 01580 753 133
www.watdon.com
(General naturalist supplies)

The Wildlife Trusts
The Kiln, Waterside, Mather Road,
Newark, Nottinghamshire NG24 1WT
Tel: 0870 0367 711
www.wildlifetrusts.org

WWT (The Wildfowl & Wetlands Trust)
Slimbridge, Gloucestershire GL2 7BT
Tel: 01453 891 900
www.wwt.org.ukAdder 101

INDEX

ACKNOWLEDGEMENTS

I would like to thank my editor at New Holland, Gareth Jones, for his patience, particularly as the deadline for the project loomed. I would also like to thank New Holland's Publishing Manager, Jo Hemmings, for giving me the chance to broaden my writing horizons beyond birds!

Also, I would like to thank the artists for their work, especially Richard Allen. Richard rose to the challenge of breathing new life into a number of mammal species when the opportunity arose, and he has shown a true understanding, and knowledge, of the mammals painted. His delightful images of Pine Marten and Badger are particular favourites, and it is fitting to see the Badger (my very favourite animal) grace the cover.

Thanks must also go to the authors of the guides that have proved invaluable as reference resources. The new generation of field guides moves on apace, and books such as the *Field Guide to the Moths of Great Britain and Ireland*, and the *Field Guide to the Dragonflies and Damselflies of Great Britain and Ireland* are exemplary examples of this new era.

The inhabitants of Greendale, with their stories as weird and wild now, as they were when I heard them first, have been with me throughout the project.

Finally, my biggest 'thank you' is to my partner Nadine. Oh, and to Dave the cat, a constant sidekick - normally sleeping by the fire, but a sidekick nonetheless!

ARTWORK ACKNOWLEDGEMENTS

Richard Allen: cover (front - top right, bottom centre; spine; back); pages 2; 5 (bottom right); 9; 145-169.

Stuart Carter/The Art Agency: cover (bottom right); pages 5 (top right); 7 (centre & bottom); 13 (top); 23-70.

Dave Daly: cover (front - top centre, top right; bottom right); pages 5 (bottom left); 7 (top); 13 (bottom); 104-144.

Sandra Doyle/The Art Agency: cover (top centre); pages 5 (top left, centre top left, centre top right, centre bottom right); 8; 14-22; 71-88; 95-103; 145; 147-8; 154-5; 158-160; 162-168.

Sandra Pond/The Art Agency: cover (bottom left); pages 10-11.

Dr Michael Roberts/The Art Agency: pages 5 (centre bottom left); 89-94.